Introduction
to
Non-Euclidean
Geometry

Harold E. Wolfe

Dover Publications, Inc.
Mineola, New York

Bibliographical Note

This Dover edition, first published 2012, is an unabridged republication of the work originally published in 1945 by Holt, Rinehart and Winston, Inc., New York.

Library of Congress Cataloging-in-Publication Data

Wolfe, Harold Eichholtz.
    Introduction to non-Eulidean geometry / Harold E. Wolfe.
        p. cm.
    Originally published: New York : Holt, Rinehart and Winston, 1945.
    Includes index.
    ISBN-13: 978-0-486-49850-8
    ISBN-10: 0-486-49850-6
    1. Geometry, Non-Euclidean. 2. Geometry, Non-Euclidean—History I. Title.

QA685.W6 2012
516.9—dc23

2012012533

Manufactured in the United States by LSC Communications
49850606   2022
www.doverpublications.com

PREFACE

This book has been written in an attempt to provide a satisfactory textbook to be used as a basis for elementary courses in Non-Euclidean Geometry. The need for such a volume, definitely intended for classroom use and containing substantial lists of exercises, has been evident for some time. It is hoped that this one will meet the requirements of those instructors who have been teaching the subject regularly, and also that its appearance will encourage others to institute such courses.

The benefits and amenities of a formal study of Non-Euclidean Geometry are generally recognized. Not only is the subject matter itself valuable and intensely fascinating, well worth the time of any student of mathematics, but there is probably no elementary course which exhibits so clearly the nature and significance of geometry and, indeed, of mathematics in general. However, a mere cursory acquaintance with the subject will not do. One must follow its development at least a little way to see how things come out, and try his hand at demonstrating propositions under circumstances such that intuition no longer serves as a guide.

For teachers and prospective teachers of geometry in the secondary schools the study of Non-Euclidean Geometry is invaluable. Without it there is strong likelihood that they will not understand the real nature of the subject they are teaching and the import of its applications to the interpretation of physical space. Among the first books on Non-Euclidean Geometry to appear in English was one, scarcely more than a pamphlet, written in 1880 by G. Chrystal. Even at that early date the value of this study for those preparing to teach was recognized. In the preface to this little brochure, Chrystal expressed his desire to bring "pangeometrical speculations under the notice of those engaged in the teaching of geometry." He wrote: "It will not be supposed that I advocate the introduction of pangeometry as a school subject; it is for the teacher that I advocate

v

such a study. It is a great mistake to suppose that it is sufficient for the teacher of an elementary subject to be just ahead of his pupils. No one can be a good elementary teacher who cannot handle his subject with the grasp of a master. Geometrical insight and wealth of geometrical ideas, either natural or acquired, are essential to a good teacher of geometry; and I know of no better way of cultivating them than by studying pangeometry."

Within recent years the number of American colleges and universities which offer courses in advanced Euclidean Geometry has increased rapidly. There is evidence that the quality of the teaching of geometry in our secondary schools has, accordingly, greatly improved. But advanced study in Euclidean Geometry is not the only requisite for the good teaching of Euclid. The study of Non-Euclidean Geometry takes its place beside it as an indispensable part of the training of a well-prepared teacher of high school geometry.

This book has been prepared primarily for students who have completed a course in calculus. However, although some mathematical maturity will be found helpful, much of it can be read profitably and with understanding by one who has completed a secondary school course in Euclidean Geometry. He need only omit Chapters V and VI, which make use of trigonometry and calculus, and the latter part of Chapter VII.

In Chapters II and III, the historical background of the subject has been treated quite fully. It has been said that no subject, when separated from its history, loses more than mathematics. This is particularly true of Non-Euclidean Geometry. The dramatic story of the efforts made throughout more than twenty centuries to prove Euclid's Parallel Postulate, culminating in the triumph of rationalism over tradition and the discovery of Non-Euclidean Geometry, is an integral part of the subject. It is an account of efforts doomed to failure, of efforts that fell short of the goal by only the scantiest margin, of errors, stupidity, discouragement and fear, and finally, of keen, penetrating insight which not only solved the particular problem, but opened up vast new and unsuspected fields of thought. It epitomizes the entire struggle of mankind for truth.

A large number of problems has been supplied — more than will be found in other books on this subject. The student will enjoy trying his hand at original exercises amid new surroundings and will find their solution a valuable discipline. But the problems are not merely practice material; they form an integral part of the book. Many important results are presented in the problems, and in some instances these results are referred to and even used.

It is believed that the material in the Appendix will be found helpful. Since most students in this country are not acquainted with the propositions of Euclid by number, we have reprinted the definitions, postulates, common notions and propositions of the First Book. An incidental contact with these propositions in their classical order may not be the least of the benefits to be derived from this study. Also included are sections on the hyperbolic functions, the theory of orthogonal circles and inversion. They are sufficiently extensive to give the reader who has not previously encountered the concepts an adequate introduction. These topics, when introduced at all in other courses, are generally presented in abstract and isolated fashion. Here they are needed and used, and the student may possibly be impressed for the first time with their practical importance.

There will be those who will classify this book as another one of those works in which the authors "wish to build up certain clearly conceived geometrical systems, and are careless of the details of the foundations on which all is to rest." It is true that no attempt has been made here to lay a complete and thoroughly rigorous foundation for either the Hyperbolic or Elliptic Geometry. We shall not be inclined to quarrel with those who contend that this should be done, for we are quite in accord with the spirit of their ideals. But experience has shown that it is best, taking into account the mathematical immaturity of those for whom this book is intended, to avoid the confusion of what would have to be, if properly done, an excessively long, and possibly repellent, preliminary period of abstract reasoning. No attempt is made to conceal the deficiency. As a matter of fact it is carefully pointed out and the way left open for the student to remove it later.

The study of Non-Euclidean Geometry is a fine, rare experience. The majority of the students entering a class in this subject come, like the geometers of old, thoroughly imbued with what is almost a reverence for Euclidean Geometry. In it they feel that they have found, in all their studies, one thing about which there can be no doubt or controversy. They have never considered the logic of its application to the interpretation of physical space; they have not even surmised that it might be a matter of logic at all. What they are told is somewhat in the nature of a shock. But the startled discomposure of the first few days is rapidly replaced during the weeks which follow by renewed confidence, an eager enthusiasm for investigation, and a greater and more substantial respect for geometry for what it really is.

Nor is this all. Here some student may understand for the first time something of the nature, significance and indispensability of postulates, not only in geometry, but in the formation of any body of reasoned doctrine. He will recognize that not everything can be proved, that something must always be taken on faith, and that the character of the superstructure depends upon the nature of the postulates in the foundation. In *South Wind*, Norman Douglas has one of his characters say, "The older I get, the more I realize that everything depends upon what a man postulates. The rest is plain sailing." Perhaps it is not too much to hope that a study of Non-Euclidean Geometry will now and then help some student to know this, and to formulate the postulates of his own philosophy consciously and wisely.

HAROLD E. WOLFE

INDIANA UNIVERSITY

C O N T E N T S

CHAPTER I. THE FOUNDATION OF EUCLIDEAN GEOMETRY

CHAPTER II. THE FIFTH POSTULATE

CHAPTER III. THE DISCOVERY OF NON-EUCLIDEAN GEOMETRY

CHAPTER IV. HYPERBOLIC PLANE GEOMETRY

CONTENTS   xi

CHAPTER VII. ELLIPTIC PLANE GEOMETRY AND TRIGONOMETRY

CHAPTER VIII. THE CONSISTENCY OF THE NON-EUCLIDEAN GEOMETRIES

APPENDIX

I. THE FOUNDATION OF EUCLIDEAN GEOMETRY

II. CIRCULAR AND HYPERBOLIC FUNCTIONS

III. THE THEORY OF ORTHOGONAL CIRCLES AND ALLIED TOPICS

IV. THE ELEMENTS OF INVERSION

I

THE FOUNDATION OF EUCLIDEAN GEOMETRY

"This book has been for nearly twenty-two centuries the encouragement
and guide of that scientific thought which is one thing with the progress
of man from a worse to a better state." — CLIFFORD

1. Introduction.

Geometry, that branch of mathematics in which are treated the
properties of figures in space, is of ancient origin. Much of its
development has been the result of efforts made throughout many
centuries to construct a body of logical doctrine for correlating the
geometrical data obtained from observation and measurement. By
the time of Euclid (about 300 b.c.) the science of geometry had
reached a well-advanced stage. From the accumulated material
Euclid compiled his *Elements*, the most remarkable textbook ever
written, one which, despite a number of grave imperfections, has
served as a model for scientific treatises for over two thousand years.

Euclid and his predecessors recognized what every student of
philosophy knows: that not everything can be proved. In building
a logical structure, one or more of the propositions must be assumed,
the others following by logical deduction. Any attempt to prove
all of the propositions must lead inevitably to the completion of a
vicious circle. In geometry these assumptions originally took the
form of postulates suggested by experience and intuition. At best
these were statements of what seemed from observation to be true or
approximately true. A geometry carefully built upon such a foundation may be expected to correlate the data of observation very
well, perhaps, but certainly not exactly. Indeed, it should be clear
that the mere change of some more-or-less doubtful postulate of

1

one geometry may lead to another geometry which, although radically different from the first, relates the same data quite as well.

We shall, in what follows, wish principally to regard geometry as an abstract science, the postulates as mere assumptions. But the practical aspects are not to be ignored. They have played no small role in the evolution of abstract geometry and a consideration of them will frequently throw light on the significance of our results and help us to determine whether these results are important or trivial.

In the next few paragraphs we shall examine briefly the foundation of Euclidean Geometry. These investigations will serve the double purpose of introducing the Non-Euclidean Geometries and of furnishing the background for a good understanding of their nature and significance.

2. The Definitions.

The figures of geometry are constructed from various elements such as points, lines, planes, curves, and surfaces. Some of these elements, as well as their relations to each other, must be left undefined, for it is futile to attempt to define all of the elements of geometry, just as it is to prove all of the propositions. The other elements and relations are then defined in terms of these fundamental ones. In laying the foundation for his geometry, Euclid[1] gave twenty-three definitions.[2] A number of these might very well have been omitted. For example, he defined a point as *that which has no part;* a line, according to him, is *breadthless length*, while a plane surface is one which *lies evenly with the straight lines on itself.* From the logical viewpoint, such definitions as these are useless. As a matter of fact, Euclid made no use of them. In modern geometries, point, line, and plane are not defined directly; they are described by being restricted to satisfy certain relations, defined or undefined, and certain postulates. One of the best of the systems constructed to

---

[1] In this book, all specific statements pertaining to Euclid's text and all quotations from Euclid are based upon T. L. Heath's excellent edition: *The Thirteen Books of Euclid's Elements*, 2nd edition (Cambridge, 1926). By permission of The Macmillan Company.

[2] These definitions are to be found in the Appendix.

serve as a logical basis for Euclidean Geometry is that of Hilbert.[3] He begins by considering three classes of *things*, points, lines, and planes. "We think of these points, straight lines, and planes," he explains, "as having certain mutual relations, which we indicate by such words as *are situated, between, parallel, congruent, continuous*, etc. The complete and exact description of these relations follows as a consequence of the axioms of geometry."

The majority of Euclid's definitions are satisfactory enough. Particular attention should be given to the twenty-third, for it will play an important part in what is to follow. It is the definition of parallel lines — the best one, viewed from an elementary standpoint, ever devised.

Parallel straight lines are straight lines which, being in the same plane and being produced indefinitely in both directions, do not meet one another in either direction.

In contrast with this definition, which is based on the concept of parallel lines *not meeting*, it seems important to call attention to two other concepts which have been used extensively since ancient times.[4] These involve the ideas that two parallel lines are lines which have the *same direction* or which are everywhere *equally distant*. Neither is satisfactory.

The *direction*-theory leads to the completion of a vicious circle. If the idea of *direction* is left undefined, there can be no test to apply to determine whether two given lines are parallel. On the other hand, any attempt to define *direction* must depend upon some knowledge of the behavior of parallels and their properties.

The *equidistant*-theory is equally unsatisfactory. It depends upon the assumption that, for the particular geometry under consideration, the locus of points equidistant from a straight line is a straight line. But this must be proved, or at least shown to be compatible with the other assumptions. Strange as it may seem, we shall shortly encounter geometries in which this is not true.

Finally, it is worth emphasizing that, according to Euclid, two

[3] *Grundlagen der Geometrie*, 7th edition (Leipzig and Berlin, 1930), or *The Foundations of Geometry*, authorized translation of the 1st edition by E. J. Townsend (Chicago, 1902). All references will be to the former unless the translation or another edition is specified. See Section 9 for this postulate system.

[4] Heath, *loc. cit.*, Vol. I, p. 190, ff.

lines in a plane *either meet or are parallel*. There is no other possible relation.

3. The Common Notions.

The ten assumptions of Euclid are divided into two sets: five are classified as *common notions*, the others as *postulates*. The distinction between them is not thoroughly clear. We do not care to go further than to remark that the common notions seem to have been regarded as assumptions acceptable to all sciences or to all intelligent people, while the postulates were considered as assumptions peculiar to the science of geometry. The five common notions are:

1. *Things which are equal to the same thing are also equal to one another.*

2. *If equals be added to equals, the wholes are equal.*

3. *If equals be subtracted from equals, the remainders are equal.*

4. *Things which coincide with one another are equal to one another.*

5. *The whole is greater than the part.*

One recognizes in these assumptions propositions of the type which at one time were so frequently described as "self-evident." From what has already been said, it should be clear that this is not the character of the assumptions of geometry at all. As a matter of fact, no "self-evident" proposition has ever been found.

4. The Postulates.

Euclid postulated the following:

1. *To draw a straight line from any point to any point.*

2. *To produce a finite straight line continuously in a straight line.*

3. *To describe a circle with any center and distance.*

4. *That all right angles are equal to one another.*

5. *That, if a straight line falling on two straight lines make the interior angles on the same side less than two right angles, the two straight lines, if produced indefinitely, meet on that side on which are the angles less than the two right angles.*

Although Euclid does not specifically say so, it seems clear that the First Postulate carries with it the idea that the line joining two points is *unique* and that two lines cannot therefore enclose a space. For example, Euclid tacitly assumed this in his proof of I, 4.[5] Like-

---

[5] The propositions of Book I are to be found, stated without proof, in the Appendix.

wise it must be inferred from the Second Postulate that the finite straight line can be produced at each extremity in only one way, so that two different straight lines cannot have a common segment. Explicit evidence of this implication first appears in the proof of XI, 1, although critical examination shows that it is needed from the very beginning of Book I. In regard to the Third Postulate, we merely remark that the word *distance* is used in place of radius, implying that each point of the circumference is at this distance from the center. The Fourth Postulate provides a standard or unit angle in terms of which other angles can be measured. Immediate use of this unit is made in Postulate 5.

The Fifth[6] Postulate plays a major role in what follows. In fact it is the starting point in the study of Non-Euclidean Geometry. One can hardly overestimate the effect which this postulate, together with the controversies which surrounded it, has had upon geometry, mathematics in general, and logic. It has been described[7] as "perhaps the most famous single utterance in the history of science." On account of its importance, we shall return to it soon and treat it at length.

5. Tacit Assumptions Made by Euclid. Superposition.

In this and the remaining sections of the chapter we wish to call attention to certain other assumptions made by Euclid. With the exception of the one concerned with superposition, they were probably made unconsciously; at any rate they were not stated and included among the common notions and postulates. These omissions constitute what is regarded by geometers as one of the gravest defects of Euclid's geometry.

Euclid uses essentially the same proof for Proposition I, 4 that is used in most modern elementary texts. There is little doubt that, in proving the congruence of two triangles having two sides and the included angle of one equal to two sides and the included angle of the other, he actually regarded one triangle as being moved in order to make it coincide with the other. But there are objections to

---

[6] This postulate is also sometimes referred to as the Eleventh or Twelfth.

[7] Keyser, *Mathematical Philosophy* (New York, 1922).

such recourse to the idea of motion without deformation in the proofs of properties of figures in space.[8] It appears that Euclid himself had no high regard for the method and used it reluctantly.

Objections arise, for example, from the standpoint that points are *positions* and are thus incapable of motion. On the other hand, if one regards geometry from the viewpoint of its application to physical space and chooses to consider the figures as capable of displacement, he must recognize that the material bodies which are encountered are always more-or-less subject to distortion and change. Nor, in this connection, may there be ignored the modern physical concept that the dimensions of bodies in motion are not the same as when they are at rest. However, in practice, it is of course possible to make an approximate comparison of certain material bodies by methods which resemble superposition. This may suggest the formulation in geometry of a postulate rendering superposition legitimate. But Euclid did not do this, although there is evidence that he may have intended Common Notion 4 to authorize the method. In answer to the objections, it also may be pointed out that what has been regarded as motion in superposition is, strictly speaking, merely a transference of attention from one figure to another.

The use of superposition can be avoided. Some modern geometers do this, for example, by *assuming* that, if two triangles have two sides and the included angle of one equal to two sides and the included angle of the other, the remaining pairs of corresponding angles are equal.[9]

6. The Infinitude of the Line.

Postulate 2, which asserts that a straight line can be produced continuously, does not necessarily imply that straight lines are infinite. However, as we shall discover directly, Euclid unconsciously assumed the infinitude of the line.

It was Riemann who first suggested the substitution of the more general postulate that the straight line is *unbounded*. In his remarkable dissertation, *Über die Hypothesen welche der Geometrie zu Grunde*

---

[8] See Heath, *loc. cit.*, Vol. I, pp. 224–228.
[9] See, for example, Hilbert, *loc. cit.*, and also Section 9.

liegen,[10] read in 1854 to the Philosophical Faculty at Göttingen, he pointed out that, however certain we may be of the unboundedness of space, we need not as a consequence infer its infinitude. He said,[11] "In the extension of space-construction to the infinitely great, we must distinguish between *unboundedness* and *infinite extent;* the former belongs to the extent relations, the latter to the measure relations. That space is an unbounded threefold manifoldness is an assumption which is developed by every conception of the outer world; according to which every instant the region of real perception is completed and the possible positions of a sought object are constructed, and which by these applications is forever confirming itself. The unboundedness of space possesses in this way a greater empirical certainty than any external experience. But its infinite extent by no means follows from this; on the other hand if we assume independence of bodies from position, and therefore ascribe to space constant curvature, it must necessarily be finite provided this curvature has ever so small a positive value."

We shall learn later that geometries, logically as sound as Euclid's, can be constructed upon the hypothesis that straight lines are boundless, being closed, but not infinite. In attempting to conceive straight lines of this character, the reader may find it helpful, provided he does not carry the analogy too far, to consider the great circles of a sphere. It is well known that in spherical geometry the great circles are *geodesics*, i.e., they are the "lines" of shortest distance between points. It will not be difficult to discover that they have many other properties analogous to those of straight lines in Euclidean Plane Geometry. On the other hand there are many striking differences. We note, for example, that these "lines," while endless, are not infinite; that, while in general two points determine a "line," two points may be so situated that an infinite number of "lines" can be drawn through them; that two "lines" always intersect in two points and enclose a space.

Even a cursory consideration of the consequences of attributing

[10] Riemann: *Gesammelte Mathematische Werke* (Leipzig, 1892).

[11] This quotation is from a translation by W. K. Clifford, in *Nature*, Vol. VIII, 1873. A translation by H. S. White is to be found in David Eugene Smith's *A Source Book in Mathematics* (New York, 1929).

to straight lines the character of being boundless, but not infinite, convinces one that Euclid tacitly assumed the infinitude of the line. One critical point at which this was done is found in the proof of I, 16. This proposition is of such importance in what follows and its consequences are so far-reaching that we present the proof here.

PROPOSITION I, 16. *In any triangle, if one of the sides be produced, the exterior angle is greater than either of the interior and opposite angles.*

Let *ABC* (Fig. 1) be the given triangle, with *BC* produced to *D*. We shall prove that
$$\angle ACD > \angle BAC.$$
Let *E* be the midpoint of *AC*. Draw *BE* and produce it to *F*, making *EF* equal to *BE*. Draw *CF*. Then triangles *BEA* and *FEC* are congruent and consequently angles *FCE* and *BAC* are equal. But
$$\angle ACD > \angle FCE.$$
Therefore
$$\angle ACD > \angle BAC.$$

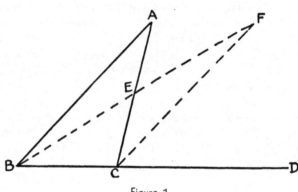

Figure 1

It should be clear now that this proof may fail, if the straight line is not infinite. As a matter of fact, the same proof can be used in spherical geometry for a spherical triangle but is valid only so long as *BF* is less than a semicircle. If *F* lies on *CD*, angle *ACD* is equal to angle *ECF* and consequently to angle *BAC*. If *BF* is greater than a semicircle, angle *ACD* will be less than angle *BAC*. Even if one conceives a geometry in which any two of the closed lines intersect in only one point, BF may be so long that F will coincide with B or lie on segment BE. In either case the proof fails. The proofs of a number of important propositions in Euclidean Geometry depend upon I, 16. Such propositions as I, 17, 18, 19, 20 and 21 will not be valid without restrictions when I, 16 does not hold.

EXERCISE

Prove Propositions I, 17, 18, 19, 20, 21.

7. Pasch's Axiom.

Another important assumption made by Euclid, without explicit statement, has been formulated by Pasch[12] as follows.

Let A, B, C be three points not lying in the same straight line and let α be a straight line lying in the plane of ABC and not passing through any of points A, B, C. Then, if the line α passes through a point of the segment AB, it will also pass through a point of the segment BC or a point of the segment AC.

It readily follows from this that, if a line enters a triangle at a vertex, it must cut the opposite side. Euclid tacitly assumed this frequently as, for example, in the proof of I, 21.

We shall make use of Pasch's Axiom many times in what follows at points where intuition cannot be depended upon to guide us as safely as it did in Euclid. In order to emphasize the importance of an explicit statement of this axiom as a characteristic of Euclidean Geometry, we remark that there are geometries in which it holds only with restrictions. It will be recognized that it is true for spherical triangles only if they are limited in size.

Pasch's Axiom is one of those assumptions classified by modern geometers as *axioms of order.*[13] These important axioms bring out the idea expressed by the word *between* and make possible an *order of sequence* of the points on a straight line.

---

[12] Pasch, *Vorlesungen über neuere Geometrie* (Berlin, 1926).
[13] See Section 9.

8. The Principle of Continuity.

One of the features of Euclid's geometry is the frequent use of constructions to prove the existence of figures having designated properties. The very first proposition is of this type and the reader will have no difficulty in recalling others. In these constructions, lines and circles are drawn, and the points of intersection of line with line, line with circle, circle with circle, are assumed to exist. Obviously, in a carefully constructed geometry, the existence of these points must be postulated or proved.

The only one of Euclid's postulates which does anything like this is the Fifth, and it applies only to a particular situation. What is needed is a postulate which will ascribe to all lines and circles that characteristic called *continuity*. This is done in a satisfactory way by one due to Dedekind.[14]

THE POSTULATE OF DEDEKIND. *If all points of a straight line fall into two classes, such that every point of the first class lies to the left of every point of the second class, then there exists one and only one point which produces this division of all points into two classes, this severing of the straight line into two portions.*

"I think I shall not err," remarks Dedekind, "in assuming that every one will at once grant the truth of this statement; the majority of my readers will be very much disappointed in learning that by this commonplace remark the secret of continuity is revealed. To this I may say that I am glad if every one finds the above principle so obvious and so in harmony with his own ideas of a line; for I am utterly unable to adduce any proof of its correctness, nor has any one the power. The assumption of this property of the line is nothing else than an axiom by which we attribute to the line its continuity, by which we find continuity in a line. If space has at all a real existence, it is *not* necessary for it to be continuous; many of its properties would be the same even were it discontinuous. And if we knew for certain that space was discontinuous there would be nothing to prevent us, in case we so desired, from filling up its gaps, in thought, and thus making it continuous; this filling up would

---

[14] Dedekind, *Essays on the Theory of Numbers*, authorized translation by W. W. Beman (Chicago, 1901), or *Gesammelte Mathematische Werke*, Vol. III, p. 322 (Brunswick, 1932).

consist in a creation of new point-individuals and would have to be effected in accordance with the above principle."

This postulate can easily be extended to cover angles and arcs as well as linear segments. As an application of the postulate we shall prove the following proposition:

The segment of line joining a point inside a circle to a point outside the circle has a point in common with the circle.

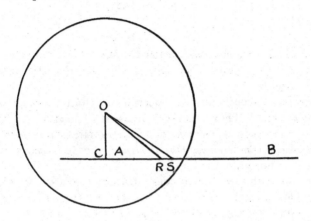

Figure 2

Let O be the center of the given circle (Fig. 2) and r its radius; let A be the point inside and B the point outside. Then

$$OA < r < OB.$$

Draw OC perpendicular to AB, produced if necessary, and note that

$$OC \leqq OA < r.$$

The points of segment AB can now be divided into two classes: those points P for which $OP < r$ and those points Q for which $OQ \geqq r$. Since, in every case,

$$OP < OQ,$$

it follows that

$$CP < CQ,$$

and thus every point P precedes (or follows) every point Q. Hence, by the Postulate of Dedekind, there exists a point R of the segment AB such that all points which precede it belong to one class and all

which follow it belong to the other class. We proceed to prove by
reductio ad absurdum that R is on the circle.

Assume that
$$OR < r$$
and choose point S on AB, between R and B, such that
$$RS < r - OR.$$
Since, in triangle ORS,
$$OS < OR + RS,$$
we conclude that
$$OS < r.$$
But this is absurd, and consequently OR cannot be less than r.

The reader can easily show in a similar way that OR cannot be
greater than r.

The idea of continuity is frequently introduced into geometry
through what is known as the *Postulate of Archimedes*. A simple,
but quite satisfactory, statement of the postulate is as follows:

*Given two linear segments, there is always some finite multiple of one
which is greater than the other.*

This can be shown[15] to be a consequence of the Postulate of Dedekind. One observes that it prescribes the exclusion of both infinite
and infinitesimal segments. It holds for arcs and angles as well as
line segments. We shall make use of it upon several occasions.

A large portion of Euclidean Geometry and also of Non-Euclidean
Geometry can be constructed without the employment of the
Principle of Continuity. We shall, however, make no particular
effort to avoid its use in what follows.

9. The Postulate System of Hilbert.

The work of such men as Pasch, Veronese, Peano and Hilbert has
placed Euclidean Geometry on a sound, logical basis. It will be
helpful to conclude this chapter by giving the system of postulates,
slightly abbreviated in form, set down by Hilbert — the system referred to in Section 2. They are arranged in six sets. It will be recalled that Hilbert begins with the undefined elements *point, line*

---

[15] See the paper by G. Vitali in Enriques' collection, *Questioni riguardanti la geometria elementare* (Bologna, 1900), or a translation into German, *Fragen der Elementar-
geometrie*, Vol. I, p. 135 (Leipzig and Berlin, 1911).

and *plane.* These elements are characterized by certain relations which are described in the postulates.

I. *The Postulates of Connection.*

1 and 2. *Two distinct points determine one and only one straight line.*

3. *There are at least two points on every line, and there are at least three points on every plane which do not lie on the same straight line.*

4 and 5. *Three points which do not lie on the same straight line determine one and only one plane.*

6. *If two points of a line lie on a plane, then all points of the line lie on the plane.*

7. *If two planes have one point in common, they have at least one other point in common.*

8. *There exist at least four points which do not lie on the same plane.*

Among the theorems to be deduced from the above set of postulates are the following:

Two distinct straight lines lying on a plane have one point or no point in common.

A line and a point not lying on that line determine a plane; so also do two distinct lines which have a point in common.

II. *The Postulates of Order.*

The postulates of this set describe an undefined relation among the points of a straight line — a relation expressed by the word *between.*

1. *If A, B and C are points of a straight line and B is between A and C, then B is also between C and A.*

2. *If A and C are two points of a straight line, there exists at least one other point of the line which lies between them.*

3. *Of any three points of a straight line, one and only one lies between the other two.*

Two points *A* and *B* determine a *segment*; *A* and *B* are the *ends* of the segment and the points between *A* and *B* are the *points of the segment*.

4. *Given three points A, B and C, which are not on the same straight line, and a straight line in the plane of ABC not passing through any of the points A, B or C, then if the line contains a point of the segment AB, it also contains a point either of the segment BC or of the segment AC.* (Pasch's Axiom.)

As deductions from the postulates already stated, we note the following:

Between any two points of a straight line there is always an unlimited number of points.

Given a finite number of points on a straight line, they can always be considered in a sequence *A*, *B*, *C*, *D*, *E*, ... , *K*, such that *B* lies between *A* and *C*, *D*, *E*, ... , *K*; *C* lies between *A*, *B* and *D*, *E*, ... , *K*; etc. There is only one other sequence with the same properties, namely, the reverse, *K*, ... , *E*, *D*, *C*, *B*, *A*.

Every straight line of a plane divides the points of the plane which are not on the line into two regions with the following properties: Every point of one region determines with every point of the other a segment containing a point of the line; on the other hand, any two points of the same region determine a segment not containing a point of the line. Thus we say that two points are on *the same side* of a line or on *opposite sides*. In a similar way, a given point of a line divides the points of a line into *half-lines* or *rays*, each ray consisting of all points of the line on *one side* of the given point.

A system of segments *AB*, *BC*, *CD*, ... , *KL* is called a *broken line* joining *A* to *L*. The points *A*, *B*, *C*, *D*, ... , *L*, as well as the points of the segments, are called the points of the broken line. If *A* and *L* coincide, the broken line is called a *polygon*. The segments are called the *sides* of the polygon, and the points *A*, *B*, *C*, *D*, ... , *K* are called the *vertices*. Polygons having 3, 4, 5, ... , *n* vertices are called, respectively, triangles, quadrangles, pentagons, ... , *n*-gons. If the vertices of a polygon are distinct and none lies on a side, and if no sides have a point in common, the polygon is called a *simple polygon.*

It follows that every simple polygon which lies in a plane divides the points of the plane not belonging to the polygon into two regions — an *interior* and an *exterior* — having the following properties: A point of one region cannot be joined to a point of the other by a broken line which does not contain a point of the polygon. Two points of the same region can, however, be so joined. The two regions can be distinguished from one another by the fact that there exist lines in the plane which lie entirely outside the polygon, but there is none which lies entirely within it.

III. *The Postulates of Congruence.*

This set of postulates introduces a new concept designated by the word *congruent.*

1. *If A and B are two points of a straight line a and A' is a point on the same or another straight line a', then there exists on a', on a given side of A', one and only one point B' such that the segment AB is congruent to the segment $A'B'$. Every segment is congruent to itself.*

2. *If a segment AB is congruent to a segment $A'B'$ and also to another segment $A''B''$, then $A'B'$ is congruent to segment $A''B''$.*

3. *If segments AB and BC of a straight line a have only point B in common, and if segments $A'B'$ and $B'C'$ of the same or another straight line a' have only B' in common, then if AB and BC are, respectively, congruent to $A'B'$ and $B'C'$, AC is congruent to $A'C'$.*

The system of two rays h and k emanating from a point O and lying on different lines is called an *angle* (h, k). The rays are called *the sides* of the angle and the point O its *vertex*. It can be proved that an angle divides the points of its plane, excluding O and the points on the sides, into two regions. Any two points of either region can always be jointed by a broken line containing neither O nor any point of either side, while no point of one region can be so joined to a point of the other. One region, called the *interior* of the angle, has the property that the segment determined by any two of its points contains only points of the region; for the other region, called the *exterior*, this is not true for every pair of points.

4. *Given an angle* (h, k) *on plane* α, *a line* a' *on the same or a different plane* α', *a point* O' *on* a', *and on line* a' *a ray* h' *emanating from* O', *then on* α' *and emanating from* O' *there is one and only one ray* k' *such that the angle* (h', k') *is congruent to angle* (h, k) *and the interior of* (h', k') *is on a given side of* a'.

5. *If the angle* (h, k) *is congruent to angle* (h', k') *and also to angle* (h'', k''), *then angle* (h', k') *is congruent to angle* (h'', k'').

The last two postulates characterize angles in the same way that III, 1 and 2 characterize segments. The final postulate of this set relates the congruence of segments and the congruence of angles.

6. *If, for triangles ABC and A'B'C', AB, AC and angle BAC are, respectively, congruent to A'B', A'C' and angle B'A'C', then the angle ABC is congruent to angle A'B'C'.*

IV. *The Postulate of Parallels.*

Given a line a *and a point* A *not lying on* a, *then there exists, in the plane determined by* a *and* A, *one and only one line which contains* A *but not any point of* a. (Playfair's Axiom.)

V. *The Postulate of Continuity.*

Given any two segments AB and CD, there always exists on the line AB a sequence of points A_1, A_2, A_3, ... , A_n, *such that the segments* AA_1, A_1A_2, A_2A_3, ... , $A_{n-1}A_n$ *are congruent to CD and B lies between A and* A_n. (Postulate of Archimedes.)

VI. *The Postulate of Linear Completeness.*

It is not possible to add, to the system of points of a line, points such that the extended system shall form a new geometry for which all of the foregoing linear postulates are valid.

Upon this foundation rests the geometry which we know as *Euclidean.*

II

THE FIFTH POSTULATE

"One of Euclid's postulates — his postulate 5 — had the fortune to be an epoch-making statement — perhaps the most famous single utterance in the history of science." — CASSIUS J. KEYSER [1]

10. Introduction.

Even a cursory examination of Book I of Euclid's Elements will reveal that it comprises three distinct parts, although Euclid did not formally separate them. There is a definite change in the character of the propositions between Proposition 26 and Proposition 27. The first twenty-six propositions deal almost entirely with the elementary theory of triangles. Beginning with Proposition 27, the middle section introduces the important theory of parallels and leads adroitly through Propositions 33 and 34 to the third part. This last section is concerned with the relations of the areas of parallelograms, triangles and squares and culminates in the famous I, 47 and its converse.

In connection with our study of the common notions and postulates we have already had occasion to examine a number of the propositions of the first of the three sections. It is a fact to be noted that the Fifth Postulate was not used by Euclid in the proof of any of these propositions. They would still be valid if the Fifth Postulate were deleted or replaced by another one compatible with the remaining postulates and common notions.

Turning our attention to the second division, consisting of Prop-

---

[1] This quotation, as well as the one at the beginning of Chapter VIII, is taken from C. J. Keyser's book, *Mathematical Philosophy*, by permission of E. P. Dutton and Company, the publishers.

ositions 27–34, we shall find it profitable to state the first three and recall their proofs.

PROPOSITION I, 27. *If a straight line falling on two straight lines make the alternate angles equal to one another, the straight lines will be parallel to one another.*

Let *ST* (Fig. 3) be a transversal cutting lines *AB* and *CD* in such a way that angles *BST* and *CTS* are equal.

Assume that *AB* and *CD* meet in a point *P* in the direction of *B* and *D*. Then, in triangle *SPT*, the exterior angle *CTS* is equal to the interior and opposite angle *TSP*. But this is impossible. It follows that *AB* and *CD* cannot meet in the direction of *B* and *D*. By similar argument, it can be shown that they cannot meet in the direction of *A* and *C*. Hence they are parallel.

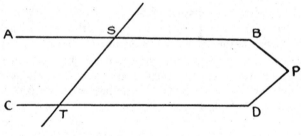

Figure 3

PROPOSITION I, 28. *If a straight line falling on two straight lines make the exterior angle equal to the interior and opposite angle on the same side, or the interior angles on the same side equal to two right angles, the straight lines will be parallel to one another.*

The proof, which follows easily from I, 27, is left to the reader.

When we come to Proposition 29, the converse of Propositions 27 and 28, we reach a critical point in the development of Euclidean Geometry. Here, for the first time, Euclid makes use of the prolix Fifth Postulate or, as it is frequently called, the *Parallel Postulate*.

PROPOSITION I, 29. *A straight line falling on parallel straight lines makes the alternate angles equal to one another, the exterior angle equal to the interior and opposite angle, and the interior angles on the same side equal to two right angles.*

Let *AB* and *CD* (Fig. 4) be parallel lines cut in points *S* and *T*, respectively, by the transversal *ST*.

Assume that angle *BST* is greater than angle *CTS*. It follows easily that the sum of angles *BST* and *STD* is greater than two right angles and consequently the sum of angles *AST* and *CTS* is less than two right angles. Then, by Postulate 5, *AB* and *CD* must meet.

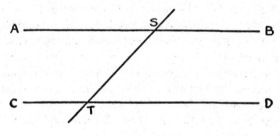

Figure 4

We conclude that angle *BST* cannot be greater than angle *CTS*. In a similar way it can be shown that angle *CTS* cannot be greater than angle *BST*. The two angles must be equal and the first part of the proposition is proved. The remaining parts are then easily verified.

There is evidence[2] that the postulates, particularly the Fifth, were formulated by Euclid, himself. At any rate, the Fifth Postulate, as such, became the target for an immediate attack upon the Elements, an attack which lasted for two thousand years. This does not seem strange when one considers, among other things, its lack of terseness when compared with the other postulates. Technically the converse of I, 17, it looks more like a proposition than a postulate and does not seem to possess to any extent that characteristic of being "self-evident." Furthermore, its tardy utilization, after so

[2] See Heath, *loc. cit.*, Vol. I, p. 202.

much had been proved without it, was enough to arouse suspicion with regard to its character.

As a consequence, innumerable attempts were made to *prove* the Postulate or eliminate it by altering the definition of parallels. Of these attempts and their failures we shall have much to recount later, for they have an all-important bearing upon our subject. For the present we wish to examine some of the substitutes for the Fifth Postulate.

11. Substitutes for the Fifth Postulate.

When, in the preceding chapter, attention was directed to the importance of the Fifth Postulate in elementary geometry and in what is to follow here, the reader may have been disturbed by an inability to recall any previous encounter with the Postulate. Such a situation is due to the fact that most writers of textbooks on geometry use some substitute postulate, essentially equivalent to the Fifth, but simpler in statement. There are many such substitutes. Heath[3] quotes nine of them. The one most commonly used is generally attributed to the geometer, Playfair, although it was stated as early as the fifth century by Proclus.

12. Playfair's Axiom.

Through a given point can be drawn only one parallel to a given line.[4]

If Playfair's Axiom is substituted for the Fifth Postulate, the latter can then be deduced as follows:

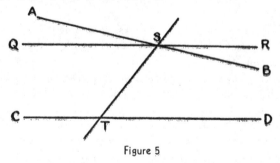

Figure 5

Given lines *AB* and *CD* (Fig. 5) cut by the transversal *ST* in such a way that the sum of angles *BST* and *DTS* is less than two right

---

[3] *Loc. cit.*, Vol. I, p. 220.
[4] That *one* parallel can always be drawn follows from I, 27 and I, 28.

angles. Construct through S the line QSR, making the sum of angles RST and DTS equal to two right angles. This line is parallel to CD by I, 28. Since lines QSR and ASB are different lines and, by Playfair's Axiom, only one line can be drawn through S parallel to CD, we conclude that AB meets CD. These lines meet in the direction of B and D, for, if they met in the opposite direction, a triangle would be formed with the sum of two angles greater than two right angles, contrary to I, 17.

Those writers of modern textbooks on geometry who prefer Playfair's Axiom to the Fifth Postulate do so because of its brevity and apparent simplicity. But it may be contended that it is neither as simple nor as satisfactory as the Postulate. C. L. Dodgson[5] points out that there is needed in geometry a practical test by which it can be proved on occasion that two lines will meet if produced. The Fifth Postulate serves this purpose and in doing so makes use of a simple geometrical picture — two finite lines cut by a transversal and having a known angular relation to that transversal. On the other hand, Playfair's Axiom makes use of the idea of parallel lines, lines which do not meet, and about the relationship of which, within the visible portion of the plane, nothing is known. Furthermore, he shows that Playfair's Axiom asserts more than the Fifth Postulate, that "all the additional assertion is superfluous and a needless strain on the faith of the learner."

EXERCISES

1. Deduce Playfair's Axiom from the Fifth Postulate.
2. Prove that each of the following statements is equivalent to Playfair's Axiom:
   (a) If a straight line intersects one of two parallel lines, it will intersect the other also.
   (b) Straight lines which are parallel to the same straight line are parallel to one another.

13. The Angle-Sum of a Triangle.

A second alternative for the Fifth Postulate is the familiar theorem:

The sum of the three angles of a triangle is always equal to two right angles.[6]

That this is a consequence of Playfair's Axiom, and hence of the

[5] *Euclid and His Modern Rivals*, pp. 40–47, 2nd edition (London, 1885).

[6] As a matter of fact, the assumption does not have to be so broad. It is sufficient to assume that there exists *one* triangle for which the angle-sum is two right angles. See Section 24.

Fifth Postulate, is well known. In order to deduce Playfair's Axiom from this assumption, we shall need two lemmas which are consequences of the assumption.

Lemma 1. An exterior angle of a triangle is equal to the sum of the two opposite and interior angles.

The proof is left to the reader.

Lemma 2. Through a given point P, there can always be drawn a line making with a given line p an angle less than any given angle α, however small.

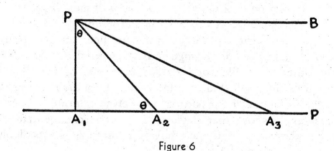

Figure 6

From P (Fig. 6) draw PA_1 perpendicular to p. Measure A_1A_2 equal to PA_1 in either direction on p and draw PA_2. Designate by θ the equal angles A_1PA_2 and A_1A_2P. Then[7]

$$\theta = \frac{\pi}{2^2}.$$

Next measure A_2A_3 equal to PA_2 and draw PA_3. Then

$$\angle A_3PA_2 = \angle A_2A_3P = \frac{\theta}{2} = \frac{\pi}{2^3}.$$

Repeated construction leads to a triangle $PA_{n-1}A_n$ for which

$$\angle A_nPA_{n-1} = \angle A_{n-1}A_nP = \frac{\pi}{2^n},$$

n being any positive integer greater than unity. By the Postulate of

---

[7] The letter π is used here to designate two right angles.

Archimedes, there exists a number k such that
$$k\alpha > \pi.$$
Then, if a positive integer n, greater than unity, is chosen such that
$$2^n > k,$$
it follows that
$$\frac{\pi}{2^n} < \alpha$$
and the lemma is proved.

We are now prepared to prove that, if the sum of the three angles of a triangle is always equal to two right angles, through any point can be drawn only one parallel to a given line.

Let P (Fig. 6) be the given point and p the given line. Draw PA_1 perpendicular to p and at P draw PB perpendicular to PA_1. By I, 28, PB is parallel to p. Consider any line through P and intersecting p, such as PA_3. Since
$$\angle PA_3A_1 + \angle A_1PA_3 = \angle BPA_3 + \angle A_1PA_3 = \frac{\pi}{2},$$
it follows that
$$\angle BPA_3 = \angle PA_3A_1.$$
Then PB is the only line through P which does not cut p, for, no matter how small an angle a line through P makes with PB, there are, by Lemma 2, always other lines through P making smaller angles with PB and cutting p, so that the first line must also cut p by the Axiom of Pasch.

14. The Existence of Similar Figures.

The following statement is also equivalent to the Fifth Postulate and may be substituted for it, leading to the same consequences.

There exists a pair of similar triangles, i.e., triangles which are not congruent, but have the three angles of one equal, respectively, to the three angles of the other.

To show that this is equivalent to the Fifth Postulate, we need only show how to deduce the latter from it, since every student of Euclid knows that the use of the Postulate leads to a geometry in which similar figures exist.

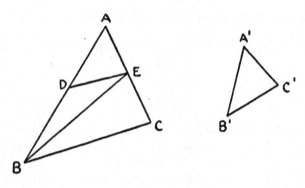

Figure 7

Given two triangles *ABC* and *A'B'C'* (Fig. 7) with angles *A*, *B* and *C* equal, respectively, to angles *A'*, *B'* and *C'*. Let *AB* be greater than *A'B'*. On *AB* construct *AD* equal to *A'B'* and on *AC* construct *AE* equal to *A'C'*. Draw *DE*. Then triangles *ADE* and *A'B'C'* are congruent. The reader can easily show that *AE* is less than *AC*, for the assumption that *AE* is greater than or equal to *AC* leads to a contradiction. It will not be difficult now to prove that the quadrilateral *BCED* has the sum of its four angles equal to four right angles.

Very shortly we shall prove,[8] without the use of the Fifth Postulate or its equivalent, that (*a*) the sum of the angles of a triangle can never be greater than two right angles, provided the straight line is assumed to be infinite, and (*b*) if one triangle has the sum of its angles equal to two right angles, then the sum of the angles of every triangle is equal to two right angles. By the use of these facts, our proof is easily completed.

By drawing *BE*, two triangles, *BDE* and *BCE*, are formed. The angle-sum for neither is greater than two right angles; if the angle-sum for either were less than two right angles, that for the other would have to be greater. We conclude that the sum of the angles for each triangle is equal to two right angles and that the same is then true for every triangle.

[8] See Section 24.

15. Equidistant Straight Lines.

Another noteworthy substitute is the following:

There exists a pair of straight lines everywhere equally distant from one another.

Once the Fifth Postulate is adopted, this statement follows, for then all parallels have this property of being everywhere equally distant. If the above statement is postulated, we can easily deduce the Fifth Postulate by first proving that there exists a triangle with the sum of its angles equal to two right angles.

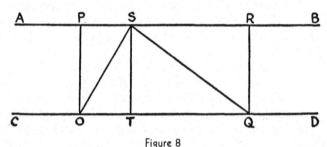

Figure 8

Let AB and CD (Fig. 8) be the two lines everywhere equally distant. From any two points O and Q on CD draw OP and QR perpendicular to AB, and from any point S on AB draw ST perpendicular to CD. By hypothesis OP, QR and ST are equal. Since right triangles OPS and OTS are congruent,

$$\angle PSO = \angle TOS.$$

Similarly

$$\angle RSQ = \angle TQS.$$

It follows that the sum of the angles of triangle OSQ is equal to two right angles.

16. Other Substitutes.

We conclude by stating without comment three other substitutes. The reader can show, in the light of later developments, that these are equivalent to the Fifth Postulate.

Given any three points not lying in a straight line, there exists a circle passing through them.

If three of the angles of a quadrilateral are right angles, then the fourth angle is also a right angle.

Through any point within an angle less than two-thirds[9] of a right angle

[9] See Section 24.

there can always be drawn a straight line which meets both sides of the angle.

These seven specimens of substitute for the Fifth Postulate are of interest as such. But they serve also to bring out the importance of the Fifth Postulate in Euclidean Geometry. Its consequences include the most familiar and most highly treasured propositions of that geometry. Without it or its equivalent there would be, for example, no Pythagorean Theorem; the whole rich theory of similar figures would disappear, and the treatment of area would have to be recast entirely. When, later on, we abandon the Postulate and replace it in turn by others which contradict it, we shall expect to find the resulting geometries strange indeed.

17. Attempts to Prove the Fifth Postulate.

We have already noted the reasons for the skepticism with which geometers, from the very beginning, viewed the Fifth Postulate as such. But the numerous and varied attempts, made throughout many centuries, to deduce it as a consequence of the other Euclidean postulates and common notions, stated or implied, all ended unsuccessfully. Before we are done we shall show why failure was inevitable. Today we know that the Postulate cannot be so derived.

But these attempts, futile in so far as the main objective was concerned, are not to be ignored. Naturally it was through them that at last the true nature and significance of the Postulate were revealed. For this reason we shall find it profitable to give brief accounts of a few of the countless efforts to prove the Fifth Postulate.

18. Ptolemy.

A large part of our information about the history of Greek geometry has come to us through the writings of the philosopher, mathematician and historian, Proclus (410–485 A.D.). He tells us that Euclid lived during the sovereignty of the first Ptolemy and that the latter himself wrote a book on the Fifth Postulate, including a proof. This must have been one of the earliest attempts to prove the Postulate. Proclus does not reproduce the proof, but from his comments we know that Ptolemy made use of the following argument in attempting to prove I, 29, without using the Postulate.

Consider two parallel lines and a transversal. The two extensions of the lines on one side of the transversal are no more parallel than their two extensions on the other side of it. Then, if the sum of the two interior angles on one side is greater than two right angles, so also is the sum of those on the other. But this is impossible, since the sum of the four angles is equal to four right angles. In a similar way it can be argued that the sum of the interior angles on one side cannot be less than two right angles. The conclusion is obvious.

19. Proclus.

Proclus himself pointed out the fallacy in the above argument by remarking that Ptolemy really assumed that through a point only one parallel can be drawn to a given line. But this is equivalent to assuming the Fifth Postulate.

Proclus submitted a proof of his own. He attempted to prove that if a straight line cuts one of two parallel lines it will cut the other also. We already know that the Fifth Postulate follows readily from this. He proceeded thus:

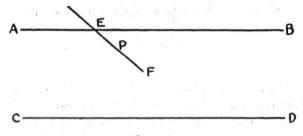

Figure 9

Given two parallel lines AB and CD (Fig. 9) with the straight line EF cutting AB at E. Assume that a point P moves along EF in the direction of F. Then the length of the perpendicular from P to AB eventually becomes greater than any length and hence greater than the distance between the parallels. Hence EF must cut CD.

The fallacy lies in the assumption that parallels are everywhere equally distant or at any rate that parallels are so related that, upon being produced indefinitely, the perpendicular from a point on one to the other remains of finite length. The former implies the Fifth Postulate, as has already been proved; the latter does also, as we shall see later.[10]

[10] See Section 47.

20. Nasiraddin.

For our next example we pass to the thirteenth century and consider the contributions of Nasiraddin (1201–1274), Persian astronomer and mathematician, who compiled an Arabic version of Euclid and wrote a treatise on the Euclidean postulates. He seems to have been the first to direct attention to the importance, in the study of the Fifth Postulate, of the theorem on the sum of the angles of a triangle. In his attempt to prove the Postulate one finds the germs of important ideas which were to be developed later.

Nasiraddin first asserted, without proof, the following:

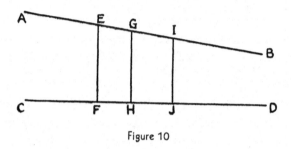

Figure 10

If two straight lines AB and CD (Fig. 10) are so related that successive perpendiculars such as EF, GH, IJ, etc., drawn to CD from points E, G, I, etc. of AB, always make unequal angles with AB, which are always acute on the side toward B, and consequently always obtuse on the side towards A, then the lines AB and CD continually diverge in the direction of A and C and, so long as they do not meet, continually converge in the direction of B and D, the perpendiculars continually growing longer in the first direction and shorter in the second. Conversely, if the perpendiculars continually become longer in the direction of A and C and shorter in the direction of B and D, the lines diverge in the first direction and converge in the other, and the perpendiculars will make with AB unequal angles, the obtuse angles all lying on the side toward A and C and the acute angles on the side towards B and D.

Next he introduced a figure destined to become famous. At the extremities of a segment *AB* (Fig. 11) he drew equal perpendiculars *AD* and *BC* on the same side, then joined *C* and *D*.

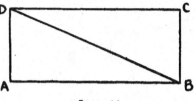

Figure 11

To prove that angles *CDA* and *DCB* are right angles, he resorted to *reductio ad absurdum*, using, without much care, the assumption stated above. Thus, if angle *DCB* were acute, *DA* would be shorter than *CB*, contrary to fact. Hence angle *DCB* is not acute. Neither is it obtuse. Of course he tacitly assumed here that, when angle *DCB* is acute, angle *CDA* must be obtuse. His argument led to the conclusion that all four angles of the quadrilateral are right angles. Then, if *DB* is drawn, the triangles *ABD* and *CDB* are congruent and the angle sum of each is equal to two right angles.

If everything were satisfactory so far, we know that the Fifth Postulate would follow easily. Nasiraddin himself presented an elaborate and exhaustive proof of this. But it is not difficult to pick flaws in the foregoing argument. For example, the assumptions made at the beginning are no more acceptable without proof than the Fifth Postulate itself. Again, when in Figure 11 it is assumed that angle *DCB* is acute, it does not follow that angle *CDA* is obtuse; as a matter of fact it will later be proved,[11] without use of the Fifth Postulate, that in such a figure these angles must be equal.

21. Wallis.

John Wallis (1616–1703) became interested in the work of Nasiraddin and described his demonstrations in a lecture at Oxford in 1651. In 1663 he offered a proof of his own. We describe it here because it is typical of those proofs which make use of an assumption equivalent to the Fifth Postulate.

Wallis suggested the assumption that, given a triangle, it is

[11] See Section 42.

possible to construct another triangle similar to it and of any size. Then he argues essentially as follows:

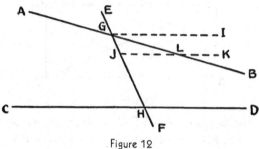

Figure 12

Given lines *AB* and *CD* (Fig. 12), cut by the transversal *EF* in points *G* and *H*, respectively, and with the sum of angles *BGH* and *DHG* less than two right angles. It is to be proved that *AB* and *CD* will meet if sufficiently produced.

It is easy to show that

$$\angle EGB > \angle GHD.$$

Then, if segment *HG* is moved along *EF*, with *HD* rigidly attached to it, until *H* coincides with the initial position of *G*, *HD* takes the position *GI*, lying entirely above *GB*. Hence, during its motion, *HD* must at some time cut *GB* as, for example, when it coincides with *JK*, cutting *GB* at *L*. Now if one constructs a triangle on base *GH* similar to triangle *GJL* — and this has been assumed to be possible — it is evident that *HD* must cut *GB*.

22. Saccheri.

In the next chapter we shall learn of the discovery of Non-Euclidean Geometry by Bolyai and Lobachewsky early in the nineteenth century. However, this discovery had all but been made by an Italian Jesuit priest almost one hundred years earlier. In 1889 there was brought to light a little book which had been published in Milan in 1733 and long since forgotten. The title of the book was *Euclides ab omni naevo vindicatus* [12] (Euclid Freed of Every Flaw),

---

[12] This book was divided into two parts, the first and more important of which is now available in English translation: Halsted, *Girolamo Saccheri's Euclides Vindicatus*, (Chicago, 1920), or see David Eugene Smith, *A Source Book in Mathematics*, p. 351 (New York, 1929).

and the author was Gerolamo Saccheri (1667–1733), Professor of Mathematics at the University of Pavia.

While teaching grammar and studying philosophy at Milan, Saccheri had read Euclid's Elements and apparently had been particularly impressed by his use of the method of *reductio ad absurdum*. This method consists of assuming, by way of hypothesis, that a proposition to be proved is false; if an absurdity results, the conclusion is reached that the original proposition is true. Later, before going to Pavia in 1697, Saccheri taught philosophy for three years at Turin. The result of these experiences was the publication of an earlier volume, a treatise on logic. In this, his *Logica demonstrativa*, the innovation was the application of the ancient, powerful method described above to the treatment of formal logic.

It was only natural that, in casting about for material to which his favorite method might be applied, Saccheri should eventually try it out on that famous and baffling problem, the proof of the Fifth Postulate. So far as we know, this was the first time anyone had thought of denying the Postulate, of substituting for it a *contradictory* statement in order to observe the consequences.

Saccheri was well prepared to undertake the task. In his *Logica demonstrativa* he had dealt ably and at length with such topics as definitions and postulates. He was acquainted with the work of others who had attempted to prove the Postulate, and had pointed out the flaws in the proofs of Nasiraddin and Wallis. As a matter of fact, it was essentially Saccheri's proof which we used above to show that the assumption of Wallis is equivalent to the Postulate.

To prepare for the application of his method, Saccheri made use of a figure with which we are already acquainted. This is the isosceles quadrilateral with the two base angles right angles.

Assuming that, in quadrilateral *ABCD* (Fig. 11), *AD* and *BC* were equal and that the angles at *A* and *B* were right angles, Saccheri easily proved, without using the Fifth Postulate or its consequences, that the angles at *C* and *D* were equal and that the line joining the midpoints of *AB* and *DC* was perpendicular to both lines. We do not reproduce his proofs here, because we shall have to give what is equivalent to them later on. Under the Euclidean hypothesis, the angles at *C* and *D* are known to be right angles. An assumption that they are acute or obtuse would imply the falsity of the Postulate. This was exactly what Saccheri's plan required. He considered three hypotheses, calling them the *hypothesis of the right angle*, the *hypothesis of the obtuse angle* and the *hypothesis of the acute angle*. Proceeding from each of the latter two assumptions, he expected to reach a contradiction. He stated and proved a number of general propositions of which the following are among the more important:

1. *If one of the hypotheses is true for a single quadrilateral, of the type under consideration, it is true for every such quadrilateral.*

2. *On the hypothesis of the right angle, the obtuse angle or the acute angle, the sum of the angles of a triangle is always equal to, greater than or less than two right angles.*

3. *If there exists a single triangle for which the sum of the angles is equal to, greater than or less than two right angles, then follows the truth of the hypothesis of the right angle, the obtuse angle or the acute angle.*

4. *Two straight lines lying in the same plane either have (even on the hypothesis of the acute angle) a common perpendicular or, if produced in the same direction, either meet one another once at a finite distance or else continually approach one another.*

Making Euclid's tacit assumption that the straight line is infinite, Saccheri had no trouble at all in disposing of the hypothesis of the obtuse angle. Upon this hypothesis he was able to prove the Fifth Postulate, which in turn implies that the sum of the angles of a triangle is equal to two right angles, contradicting the hypothesis. It will be seen later, however, that if he had not assumed the infinitude of the line, as he did in making use of Euclid I, 18 in his argument, the contradiction could never have been reached.

But the hypothesis of the acute angle proved more difficult. The expected contradiction did not come. As a matter of fact, after a long sequence of propositions, corollaries and scholia, many of which were to become classical theorems in Non-Euclidean Geometry, Saccheri concluded lamely that the hypothesis leads to the absurdity that there exist two straight lines which, when produced to infinity, merge into one straight line and have a common perpendicular at infinity. One feels very sure that Saccheri himself was not thoroughly convinced by a demonstration involving such hazy concepts. Indeed, it is significant that he tried a second proof,

though with no greater success. Had Saccheri suspected that he had
reached no contradiction simply because there was none to be
reached, the discovery of Non-Euclidean Geometry would have been
made almost a century earlier than it was. Nevertheless, his is
really a remarkable work. If the weak ending is ignored, together
with a few other defects, the remainder marks Saccheri as a man
who possessed geometric skill and logical penetration of high order.
It was he who first had a glimpse of the three geometries, though he
did not know it. He has been aptly compared with his fellow-
countryman, Columbus, who went forth to discover a new route to
a known land, but ended by discovering a new world.

23. Lambert.

In Germany, a little later, Johann Heinrich Lambert (1718–1777)
also came close to the discovery of Non-Euclidean Geometry. His
investigations on the theory of parallels were stimulated by a dissertation by Georgius Simon Klügel which appeared in 1763. It
appears that Klügel was the first to express some doubt about the
possibility of proving the Fifth Postulate.

There is a striking resemblance between Saccheri's *Euclides Vindicatus* and Lambert's *Theorie der Parallellinien*,[13] which was written in
1766, but appeared posthumously. Lambert chose for his fundamental figure a quadrilateral with three right angles, that is, one-half
the isosceles quadrilateral used by Saccheri. He proposed three
hypotheses in which the fourth angle of this quadrilateral was in
turn right, obtuse and acute. In deducing propositions under the
second and third hypotheses, he was able to go much further than
Saccheri. He actually proved that the area of a triangle is proportional to the difference between the sum of its angles and two
right angles, to the *excess* in the case of the second hypothesis and to
the *deficit* in the case of the third. He noted the resemblance of the
geometry based on the second hypothesis to spherical geometry in
which the area of a triangle is proportional to its spherical excess,
and was bold enough to lean toward the conclusion that in a like
manner the geometry based on the third hypothesis could be verified

---

[13] This tract, as well as Saccheri's treatise, is reproduced in Engel and Stäckel,
Die Theorie der Parallellinien von Euklid bis auf Gauss (Leipzig, 1895).

on a sphere with imaginary radius. He even remarked that in the third case there is an absolute unit of length.

He, like Saccheri, was able to rule out the geometry of the second hypothesis, but he made the same tacit assumptions without which no contradictions would have been reached. His final conclusions for the third geometry were indefinite and unsatisfactory. He seemed to realize that the arguments against it were largely the results of tradition and sentiment. They were, as he said, *argumenta ab amore et invidia ducta*, arguments of a kind which must be banished altogether from geometry, as from all science.

One cannot fail to note that, while geometers at this time were still attempting to prove the Postulate, nevertheless they were attacking the problem with more open minds. The change had been slow, but there is no doubt that old prejudices were beginning to disappear. The time was almost ripe for far-reaching discoveries to be made.

24. Legendre.

Finally, we must not fail to include, in our discussion of the attempts to prove the Postulate, some account of the extensive writings of Adrien Marie Legendre (1752–1833). Not that he made any valuable original contribution to the subject, for most of his results had already been obtained substantially by his predecessors. But the simple, straightforward style of his proofs brought him a large following and helped to create an interest in these ideas just at a time when geometers were on the threshold of great discoveries. Some of his proofs, on account of their elegance, are of permanent value.

His attack upon the problem was much like Saccheri's and the results which he obtained were to a large extent the same. He chose, however, to place emphasis upon the angle-sum of the triangle and proposed three hypotheses in which the sum of the angles was, in turn, equal to, greater than and less than two right angles, hoping to be able to reject the last two.

Unconsciously assuming the straight line infinite, he was able to eliminate the geometry based on the second hypothesis by proving the following theorem:

The sum of the three angles of a triangle cannot be greater than two right angles.

Assume that the sum of the angles of a triangle ABC (Fig. 13) is $180° + \alpha$ and that angle CAB is not greater than either of the others.

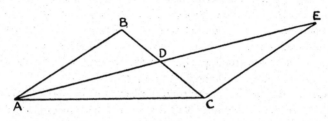

Figure 13

Join A to D, the midpoint of BC, and produce AD to E so that DE is equal to AD. Draw CE. Then triangles BDA and CDE are congruent. It follows easily that the sum of the angles of triangle AEC is equal to the sum of the angles of triangle ABC, namely to $180° + \alpha$, and that one of the angles CAE and CEA is equal to or less than one-half angle CAB. By applying the same process to triangle AEC, one obtains a third triangle with angle-sum equal to $180° + \alpha$ and one of its angles equal to or less than $\frac{1}{2^2} \angle CAB$. When this construction has been made n times, a triangle is reached which has the sum of its angles equal to $180° + \alpha$ and one of its angles equal to or less than $\frac{1}{2^n} \angle CAB$.

By the Postulate of Archimedes, we know that there is a finite multiple of α, however small α may be, which exceeds angle CAB, i.e.,

$$\angle CAB < k\alpha.$$

If n is chosen so large that

$$k < 2^n,$$

then

$$\frac{1}{2^n} \angle CAB < \alpha,$$

and the sum of *two* of the angles of the triangle last obtained is greater than two right angles. But that is impossible.

One recognizes at once the similarity of this proof to that of

Euclid I, 16. Here also one sees how important for the proof is the assumption of the infinitude of the line.

But, although he made numerous attempts, Legendre could not dispose of the third hypothesis. This, as Gauss remarked, was the reef on which all the wrecks occurred. We know now that these efforts were bound to be futile. It will be of interest, however, to examine one of his attempted proofs that the sum of the angles of a triangle cannot be less than two right angles.

Assume that the sum of the three angles of triangle ABC (Fig. 14) is $180° - \alpha$ and that angle BAC is not greater than either of the others.

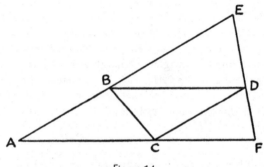

Figure 14

Construct on side BC a triangle BCD congruent to triangle ABC, with angles DBC and DCB equal, respectively, to angles BCA and CBA. Then draw through D any line which cuts AB and AC produced in E and F, respectively.

The sum of the angles of triangle BCD is also $180° - \alpha$. Since, as proved above, the sum of the angles of a triangle cannot be greater than two right angles, the sum of the angles of triangle BDE and also of triangle CDF cannot be greater than $180°$. Then the sum of all of the angles of all four triangles cannot be greater than $720° - 2\alpha$. It follows that the sum of the three angles of triangle AEF cannot be greater than $180° - 2\alpha$.

If this construction is repeated until n such triangles have been formed in turn, the last one will have its angle sum not greater than $180° - 2^n\alpha$. But, since a finite multiple of α can be found which is greater than two right angles, n can be chosen so large that a triangle will be reached which has the sum of its angles negative, and this is absurd.

The fallacy in this proof lies in the assumption that, through any point within an angle less than two-thirds of a right angle, there can always be drawn a straight line which meets both sides of the angle. This is equivalent, as we have already remarked, to the assumption of the Fifth Postulate.

The proofs of the following sequence of important theorems are essentially those of Legendre.

If the sum of the angles of a triangle is equal to two right angles, the same is true for all triangles obtained from it by drawing lines through vertices to points on the opposite sides.

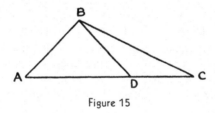

Figure 15

If the sum of the angles of triangle ABC (Fig. 15) is equal to two right angles, then the same must be true for triangle ABD, one of the two triangles into which triangle ABC is subdivided by the line joining vertex B to point D on the opposite side. For the sum of the angles of triangle ABD cannot be greater than two right angles (as proved above, with the tacit assumption of the infinitude of the straight line), and if the sum were less than two right angles, that for triangle BDC would have to be greater than two right angles.

If there exists a triangle with the sum of its angles equal to two right angles, an isosceles right triangle can be constructed with the sum of its angles equal to two right angles and the legs greater in length than any given line segment.

Let the sum of the angles of triangle *ABC* (Fig. 16) be equal to two right angles. If *ABC* is not an isosceles right triangle, such a triangle, with the sum of its angles equal to two right angles, can be constructed by drawing altitude *BD* and then, if neither of the

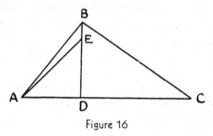

Figure 16

resulting right triangles is isosceles, measuring off on the longer leg of one of them a segment equal to the shorter. For example, if *BD* is greater than *AD*, measure *DE* equal to *AD* and draw *AE*.

If two such isosceles right triangles which are congruent are adjoined in such a way that the hypotenuse of one coincides with that of the other, a quadrilateral will be formed with its angles all right angles and its sides equal. With four such congruent quadrilaterals there can be formed another of the same type with its sides twice as long as those of the one first obtained. If this construction is repeated often enough, one eventually obtains, after a finite number of operations, a quadrilateral of this kind with its sides greater than any given line segment. A diagonal of this quadrilateral divides it into two right triangles of the kind described in the theorem.

If there exists a single triangle with the sum of its angles equal to two right angles, then the sum of the angles of every triangle will be equal to two right angles.

Given a triangle with the sum of its three angles equal to two right angles, it is to be proved that any other triangle *ABC* has its angle sum equal to two right angles. It may be assumed that *ABC*

(Fig. 17) is a right triangle, since any triangle can be divided into

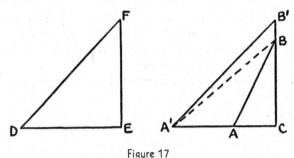

Figure 17

two right triangles. By the preceding theorem, there can be constructed an isosceles right triangle DEF, with the sum of its three angles equal to two right angles and its equal legs greater than the legs of triangle ABC. Produce CA and CB to A' and B', respectively, so that $CA' = CB' = ED = EF$, and join A' to B and to B'. Since triangles $A'CB'$ and DEF are congruent, the former has the sum of its angles equal to two right angles and the same is true for triangle $A'BC$ and finally for ABC.

As an immediate consequence of these results, Legendre obtained the theorem:

If there exists a single triangle with the sum of its angles less than two right angles, then the sum of the angles of every triangle will be less than two right angles.

25. Some Fallacies in Attempts to Prove the Postulate.

Of the so-called proofs of the Fifth Postulate already considered, some have depended upon the conscious or unconscious use of a substitute, equivalent to the Postulate in essence, and have thus begged the question. Others have made use of the *reductio ad absurdum* method, but in each case with results which have been nebulous and unconvincing. But there are other types of attempted proof. Some of them are very ingenious and seem quite plausible, with fallacies which are not easy to locate. We shall conclude this chapter by examining two of them.

26. The Rotation Proof.

This ostensible proof, due to Bernhard Friedrich Thibaut[14] (1775–1832) is worthy of note because it has from time to time appeared in elementary texts and has otherwise been indorsed. The substance of the proof is as follows:
In triangle ABC (Fig. 18), allow side AB to rotate about A, clock-

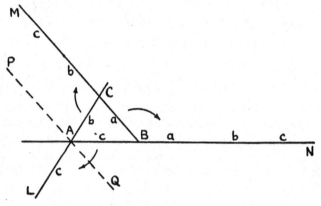

Figure 18

wise, until it coincides with CA produced to L. Let CL rotate clockwise about C until it coincides with BC produced to M. Finally, when BM has been rotated clockwise about B, until it coincides with AB produced to N, it appears that AB has undergone a complete rotation through four right angles. But the three angles of rotation are the three exterior angles of the triangle, and since their sum is equal to four right angles, the sum of the interior angles must be equal to two right angles.

[14] *Grundriss der reinen Mathematik*, 2nd edition (Göttingen, 1809).

This proof is typical of those which depend upon the idea of direction. The circumspect reader will observe that the rotations take place about different points on the rotating line, so that not only rotation, but translation, is involved. In fact, one sees that the segment AB, after the rotations described, has finally been translated along AB through a distance equal to the perimeter of the triangle. Thus it is assumed in the proof that the translations and rotations are independent, and that the translations may be ignored. But this is only true in Euclidean Geometry and its assumption amounts to taking for granted the Fifth Postulate. The very same argument can be used for a spherical triangle, with the same conclusion, although the sum of the angles of any such triangle is always greater than two right angles.

The proof does not become any more satisfactory if one attempts to make the rotations about a single point, say A. For if PQ is drawn through A, making angle PAL equal to angle MCA, one must not conclude that angle PAB will equal angle CBN. This would, as Gauss[15] pointed out, be equivalent to the assumption that if two straight lines intersect two given lines and make equal corresponding angles with one of them, then they must make equal corresponding angles with the other also. But this will be recognized as essentially the proposition to be proved. For if two straight lines make equal corresponding angles with a third, they are parallel by Euclid I, 28. To conclude that they make equal angles with any other line which intersects them amounts to the assumption of I, 29.

27. Comparison of Infinite Areas.

Another proof, which has from time to time captured the favor of the unwary, is due to the Swiss mathematician, Louis Bertrand[16]

---

[15] See his correspondence with Schumacher, Engel and Stäckel, *loc. cit.*, pp. 227-230.

[16] Développement nouveau de la partie élémentaire des Mathématiques, Vol. II, p. 19 (Geneva, 1778).

(1731–1812). He attempted to prove the Fifth Postulate directly, using in essence the following argument:

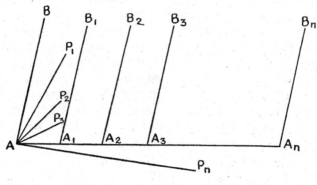

Figure 19

Given two lines AP_1 and A_1B_1 (Fig. 19) cut by the transversal AA_1 in such a way that the sum of angles P_1AA_1 and AA_1B_1 is less than two right angles, it is to be proved that AP_1 and A_1B_1 meet if sufficiently produced.

Construct AB so that angle BAA_1 is equal to angle $B_1A_1A_2$, where A_2 is a point on AA_1 produced through A_1. Then AP_1 will lie within angle BAA_1, since angle P_1AA_1 is less than angle $B_1A_1A_2$. Construct AP_2, AP_3, ... , AP_n so that angles P_1AP_2, P_2AP_3, ... , $P_{n-1}AP_n$ are all equal to angle BAP_1. Since an integral multiple of angle BAP_1 can be found which exceeds angle BAA_1, n can be chosen so large that AP_n will fall below AA_1 and angle BAP_n be greater than angle BAA_1. Since the infinite sectors BAP_1, P_1AP_2, ... , $P_{n-1}AP_n$ can be superposed, they have equal areas and each has an area equal to that of the infinite sector BAP_n divided by n.

Next, on AA_1 produced through A_1, measure A_1A_2, A_2A_3, ... , $A_{n-1}A_n$ all equal to AA_1, and construct A_2B_2, A_3B_3, ... , A_nB_n so that they make with AA_n the same angle which A_1B_1 makes with that line. Then the infinite strips BAA_1B_1, $B_1A_1A_2B_2$, ... , $B_{n-1}A_{n-1}A_nB_n$ can be superposed and thus have equal areas, each equal to the area of the infinite strip BAA_nB_n divided by n. Since the infinite sector BAP_n includes the infinite strip BAA_nB_n, it follows that the area of the sector BAP_1 is greater than that of the strip

BAA_1B_1, and therefore AP_1 must intersect A_1B_1 if produced sufficiently far.

The fallacy lies in treating infinite magnitudes as though they were finite. In the first place, the idea of congruence as used above for infinite areas has been slurred over and not even defined. Again, one should note that reasoning which is sound for finite areas need not hold for those which are infinite. In order to emphasize the weakness of the proof, one may compare, using the same viewpoint, the areas of the infinite sectors BAA_n and $B_1A_1A_n$. Since these sectors can be superposed, one might as a consequence conclude that they have equal areas. On the other hand, the former appears to be larger than the latter and to differ from it by the area of the infinite strip BAA_1B_1. As a matter of fact, any comparison of infinite magnitudes must ultimately be made to depend upon the process of finding the limit of a fraction, both the numerator and the denominator of which become infinite.

THE DISCOVERY OF NON-EUCLIDEAN GEOMETRY

"Out of nothing I have created a strange new universe."

— JOHANN BOLYAI

28. Introduction.

The beginning of the nineteenth century found the obstinate puzzle of the Fifth Postulate still unsolved. But one should not get the impression that the efforts to prove the Postulate, made throughout more than twenty centuries, were altogether fruitless. Slowly but surely they had directed the speculations of geometers to the point where the discovery of Non-Euclidean Geometry could not long be delayed. In retrospect, one wonders at first that this preparation should have taken so long, but on second thought marvels that such a momentous discovery came as early as it did.

At the time the new ideas were crystalizing, the philosophy of Kant (1724–1804) dominated the situation, and this philosophy treated space not as empirical, but as intuitive. From this viewpoint, space was regarded as something already existing in the mind and not as a concept resulting from external experience. In that day it required not only perspicacity, but courage, to recognize that geometry becomes an experimental science, once it is applied to physical space, and that its postulates and their consequences need only be accepted if convenient and if they agree reasonably well with experimental data.

But the change of viewpoint gradually came. The discovery of Non-Euclidean Geometry led eventually to the complete destruction of the Kantian space conception and at last revealed not only the true distinction between concept and experience but, what is even more important, their interrelation.

We are not surprised that, when the time came, the discovery of Non-Euclidean Geometry was not made by one man, but independently by several in different parts of the world. This has happened more than once in the history of mathematics and it will doubtless happen again. The father of Johann Bolyai, one of the founders of Non-Euclidean Geometry, predicted this when, in a letter to his son urging that he make public his discoveries without delay, he wrote,[1] "It seems to me advisable, if you have actually succeeded in obtaining a solution of the problem, that, for a two-fold reason, its publication be hastened: first, because ideas easily pass from one to another who, in that case, can publish them; secondly, because it seems to be true that many things have, as it were, an epoch in which they are discovered in several places simultaneously, just as the violets appear on all sides in springtime."

And so it happened that independently and at about the same time the discovery of a logically consistent geometry, in which the Fifth Postulate was denied, was made by Gauss in Germany, Bolyai in Hungary and Lobachewsky in Russia.

29. Gauss.

At the turn of the century, during those critical years in the evolution of geometry, the dominant figure in the mathematical world was Carl Friedrich Gauss (1777–1855). Naturally he took no small part in the development of the ideas which led to the discovery of the new systems of geometry. Few of the results of his many years of meditation and research on the problems associated with the Fifth Postulate were published or made public during his lifetime. Some letters written to others interested in those problems, two published reviews of certain tracts on parallels and a few notes discovered among his papers furnish meager but sufficient evidence that he was probably the first to understand clearly the possibility of a logically sound geometry different from Euclid's. It was he who

[1] Paul Stäckel: *Wolfgang und Johann Bolyai, Geometrische Untersuchungen*, Vol. I, p. 86 (Leipzig and Berlin, 1913).

first called the new geometry *Non-Euclidean*. The correspondence and reviews[2] referred to outline rather clearly the progress he made in the study of parallels, and show that recognition of the new geometry did not come suddenly but only after many years of thought. It seems clear that, even as late as the first decade of the new century, Gauss, traveling in the footsteps of Saccheri and Lambert, with whose books he may have been familiar, was still attempting to prove the Fifth Postulate by the *reductio ad absurdum* method, but that he fully recognized the profound character of the obstacles encountered. It was during the second decade that he began the formulation of the idea of a new geometry, to develop the elementary theorems and to dispel his doubts. No words can describe the nature of his discoveries, the significance he attached to them, his attitude toward the current concept of space and his fear of being misunderstood, half so well as his own words in a letter written at Göttingen on November 8, 1824 to F. A. Taurinus. The following is a translation of this important document.[3] "I have not read without pleasure your kind letter of October 30th with the enclosed abstract, all the more because until now I have been accustomed to find little trace of real geometrical insight among the majority of people who essay anew to investigate the so-called Theory of Parallels.

"In regard to your attempt, I have nothing (or not much) to say except that it is incomplete. It is true that your demonstration of the proof that the sum of the three angles of a plane triangle cannot be greater than 180° is somewhat lacking in geometrical rigor. But this in itself can easily be remedied, and there is no doubt that the impossibility can be proved most rigorously. But the situation is quite different in the second part, that the sum of the angles cannot be less than 180°; this is the critical point, the reef on which all the wrecks occur. I imagine that this problem has not engaged you very long. I have pondered it for over thirty years, and I do not believe that anyone can have given more thought to this second part than I, though I have never published anything on it. The assumption that the sum of the three angles is less than 180° leads to a

[2] Engel and Stäckel have collected some of these in their sourcebook, *Die Theorie der Parallellinien von Euklid bis auf Gauss* (Leipzig, 1895).
[3] For a photographic facsimile of this letter see Engel and Stäckel, *loc. cit.*

curious geometry, quite different from ours (the Euclidean), but thoroughly consistent, which I have developed to my entire satisfaction, so that I can solve every problem in it with the exception of the determination of a constant, which cannot be designated *a priori*. The greater one takes this constant, the nearer one comes to Euclidean Geometry, and when it is chosen infinitely large the two coincide. The theorems of this geometry appear to be paradoxical and, to the uninitiated, absurd; but calm, steady reflection reveals that they contain nothing at all impossible. For example, the three angles of a triangle become as small as one wishes, if only the sides are taken large enough; yet the area of the triangle can never exceed a definite limit, regardless of how great the sides are taken, nor indeed can it ever reach it. All my efforts to discover a contradiction, an inconsistency, in this Non-Euclidean Geometry have been without success, and the one thing in it which is opposed to our conceptions is that, if it were true, there must exist in space a linear magnitude, *determined for itself* (but unknown to us). But it seems to me that we know, despite the say-nothing word-wisdom of the metaphysicians, too little, or too nearly nothing at all, about the true nature of space, to consider as *absolutely impossible* that which appears to us unnatural. If this Non-Euclidean Geometry were true, and it were possible to compare that constant with such magnitudes as we encounter in our measurements on the earth and in the heavens, it could then be determined *a posteriori*. Consequently in jest I have sometimes expressed the wish that the Euclidean Geometry were not true, since then we would have *a priori* an absolute standard of measure.

"I do not fear that any man who has shown that he possesses a thoughtful mathematical mind will misunderstand what has been said above, but in any case consider it a private communication of which no public use or use leading in any way to publicity is to be made. Perhaps I shall myself, if I have at some future time more leisure than in my present circumstances, make public my investigations."

The failure of Gauss to make public his results made it inevitable that the world withhold a portion of the honor which might have been entirely his. As we shall see, others who came to the same conclusions, although probably a little later, promptly extended the ideas and courageously published them. To accord to these the glory in its fullest form is only just. But one cannot fail altogether to sympathize with Gauss in his reluctance to divulge his discoveries. By his day many prominent mathematicians, dominated by the philosophy of Kant, had come to the conclusion that the mystery of the Fifth Postulate could never be solved. There were still those who continued their investigations, but they were likely to be regarded as cranks. It was probably the derision of smug and shallow-minded geometers that Gauss feared. Nor can one safely say that he had less courage than those who made public their results. Compared to him, they were obscure, with no reputations to uphold and nothing much to lose. Gauss, on the other hand, had climbed high. If he fell, he had much farther to fall.

In a letter[4] to Schumacher, dated May 17, 1831, and referring to the problem of parallels, Gauss wrote: "I have begun to write down during the last few weeks some of my own meditations, a part of which I have never previously put in writing, so that already I have had to think it all through anew three or four times. But I wished this not to perish with me."

Consequently, among his papers there is to be found a brief account of the elementary theory of parallels for the new geometry. We have already noted that one of the simplest substitutes for the Fifth Postulate is the so-called Playfair Axiom. In rejecting the Postulate Gauss, like Bolyai and Lobachewsky, chose to assume that through a point more than one parallel (in the sense of Euclid) can be drawn to a given line.

There is no need to sketch through the details of what little he jotted down; it is essentially similar to the elementary theory presented in the first few pages of the next chapter. He did not go far in recording his meditations; his notes came to a sudden halt. For on February 14, 1832 he received a copy of the famous *Appendix* by Johann Bolyai.

30. Bolyai.

While studying at Göttingen, Gauss numbered among his friends a Hungarian, Wolfgang Bolyai[5] (Bolyai Farkas, 1775–1856), who

---

[4] See Engel and Stäckel, *loc. cit.*, p. 230.
[5] Pronounced Bol'yah-eh.

was a student there from 1796 to 1799. It is quite certain that the two frequently discussed problems related to the theory of parallels. After they left the University, they continued their intercourse by correspondence. A letter[6] written by Gauss to Bolyai in 1799 shows that both were at that time still attempting to prove the Fifth Postulate. In 1804, Bolyai, convinced that he had succeeded in doing this, presented his ideas in a little tract entitled *Theoria Parallelarum*,[7] which he sent to Gauss, enclosed with a letter. But the proof was incorrect, and Gauss, in replying, pointed out the error. Undaunted, Bolyai continued to reason along the same lines and, four years later, sent to Gauss a supplementary paper.[8] He apparently became discouraged when Gauss did not reply, and turned his attention to other matters. However, during the next two decades, despite varied interests as professor, poet, dramatist, musician, inventor and forester, he managed to collect his ideas on elementary mathematics and finally publish them in 1832–33 in a two volume work which we shall call briefly the *Tentamen*.[9] Wolfgang Bolyai was a talented and capable man, but his claim to fame must doubtless be based upon the fact that he was the father of Johann.

For on December 15, 1802 was born Johann Bolyai (Bolyai Janos, 1802–1860). "He is, Heaven be praised," wrote Wolfgang to Gauss in 1803, "a healthy and very beautiful child, with a good disposition, black hair and eyebrows, and burning deep blue eyes, which at times sparkle like two jewels." And during those years leading up to the publication of the *Tentamen*, Johann had been growing to manhood.

His father gave him his early instruction in mathematics, so that it does not seem unnatural that he should have become interested in the theory of parallels. Nor is it a matter for surprise to learn that, by the time he had become a student in the Royal College for Engineers at Vienna in 1817, he had devoted much thought to the

---

[6] Engel and Stäckel, *loc. cit.*, p. 219.

[7] See Stäckel, *Wolfgang und Johann Bolyai*, Vol. II, p. 5, for a German translation. The original was in Latin.

[8] Stäckel, *loc. cit.*, Vol. II, p. 16.

[9] The title is: *Tentamen juventutem studiosam in elementa matheseos purae, elementaris ac sublimioris, methodo intuitiva, evidentiaque huic propria, introducendi. Cum appendice triplice.* See Stäckel, *loc. cit.*, Vol. II, p. 25.

problem of the proof of the Fifth Postulate, despite the fact that his father, recalling his own unsuccessful efforts, recommended that the ancient enigma was something to be left entirely alone. But, by 1820, his efforts to prove the Postulate by the substitution of a contradictory assumption began to yield results of a different nature. His attention was gradually directed toward the possibility of formulating a general geometry, an *Absolute Science of Space*, with Euclidean Geometry as a special case.

In his attempts to prove the Fifth Postulate by denying it, Bolyai chose to regard that assumption in the form which we have already designated as Playfair's Axiom, and which asserts that one and only one parallel line can be drawn through a given point to a given line. The denial of the Postulate then implies either that *no* parallel to the line can be drawn through the point or that *more than one* such parallel can be drawn. But, as a consequence of Euclid I, 27 and 28, provided the straight line is regarded as infinite, the former of the two implications must be discarded. Furthermore, if there are at least two parallels to the line through the point, then there must be an infinite number of parallels in the sense of Euclid. If, for example, the two lines *CD* and *EF* (Fig. 20) through *P* do not cut *AB*, then the same will be true for all lines through *P* which lie within the vertical angles *EPC* and *DPF*. In substance Bolyai, as did Gauss

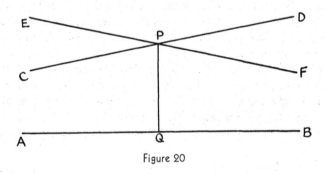

Figure 20

and Lobachewsky, then argued that if one starts with *PQ* perpendicular to *AB* and allows *PQ* to rotate about *P* in either direction, it will continue to cut *AB* awhile and then cease to cut it. He was thus led to postulate the existence of two lines through *P* which separate the lines which cut *AB* from those which do not. Since for rotation of *PQ* in either direction there is no last cutting line, these postulated lines must be the first of the non-cutting lines. It will develop that these two lines parallel to *AB* have properties quite different from the other lines through *P* which do not cut *AB*.

The results which followed as a consequence of these assumptions aroused the greatest wonder in the young Bolyai. As the geometry developed and no contradictions appeared, this wonder grew and he began to feel something of the significance of what he was doing. What seemed to impress him most were the propositions which did not depend upon any parallel postulate at all, but which were common to all geometries regardless of what assumptions were made about parallels. These he regarded as stating absolute facts about space and forming the basis of an absolute geometry.

These ideas had certainly begun to take form, however vaguely, by 1823 when Bolyai was only twenty-one years old. The following extract from a letter[10], written to his father on November 3, 1823, shows how far he had gone with his discoveries and how deeply he was affected by them.

"It is now my definite plan to publish a work on parallels as soon as I can complete and arrange the material and an opportunity presents itself; at the moment I still do not clearly see my way through, but the path which I have followed gives positive evidence that the goal will be reached, if it is at all possible; I have not quite reached it, but I have discovered such wonderful things that I was amazed and it would be an everlasting piece of bad fortune if they were lost. When you, my dear Father, see them, you will understand; at present I can say nothing except this: that *out of nothing I have created a strange new universe*. All that I have sent you previously is like a house of cards in comparison with a tower. I am no less convinced that these discoveries will bring me honor, than I would be if they were completed."

In reply, the elder Bolyai suggested that the proposed work be published as an appendix to his *Tentamen*, and urged that this be done with as little delay as possible.[11] But the formulation of re-

[10] Stäckel, *loc. cit.*, Vol. I, p. 85.   [11] See article 28.

sults and the expansion of ideas came slowly. In February, 1825, however, Johann visited his father and brought along an outline of his work. Finally in 1829 he submitted his manuscript and, despite the fact that father and son disagreed on a few points, there was published in 1832 the *Appendix*.[12]

Previously, in 1831, eager to know what Gauss would have to say about his son's discoveries, Wolfgang had sent him an abridgment of the Appendix, but it failed to reach him. In February, 1832, Gauss received an advance copy of the *Appendix*. His response,[13] written to Wolfgang on March 6, 1832, contains the following remarks about the work of Johann.

"If I begin with the statement that I dare not praise such a work, you will of course be startled for a moment: but I cannot do otherwise; to praise it would amount to praising myself; for the entire content of the work, the path which your son has taken, the results to which he is led, coincide almost exactly with my own meditations which have occupied my mind for from thirty to thirty-five years. On this account I find myself surprised to the extreme.

"My intention was, in regard to my own work, of which very little up to the present has been published, not to allow it to become known during my lifetime. Most people have not the insight to understand our conclusions and I have encountered only a few who received with any particular interest what I communicated to them. In order to understand these things, one must first have a keen perception of what is needed, and upon this point the majority are quite confused. On the other hand it was my plan to put all down on paper eventually, so that at least it would not finally perish with me.

"So I am greatly surprised to be spared this effort, and am overjoyed that it happens to be the son of my old friend who outstrips me in such a remarkable way."

---

[12] The complete title is: *Appendix. Scientiam Spatii absolute veram exhibens: a veritate aut falsitate Axiomatis XI Euclidei (a priori haud umquam decidenda) independentem: adjecta ad casum falsitatis, quadratura circuli geometrica.* See the German translation by Stäckel, *loc. cit.*, Vol. II, p, 183, the English translation by G. B. Halsted, 4th edition (Austin, Texas, 1896), or David Eugene Smith: *A Source Book in Mathematics*, p. 375 (New York, 1929).

[13] Stäckel, *loc. cit.*, Vol. I, p. 92.

When Johann received a copy of this letter from his father he was far from elated. Instead of the eulogies which he had anticipated, it brought him, in his opinion, only the news that another had made the same discoveries independently and possibly earlier. He even went so far as to suspect that, before the *Appendix* was completed, his father had confided some of his ideas to Gauss, who in turn had appropriated them for his own use. These suspicions were eventually dispelled, but Johann never felt that Gauss had accorded him the honor that was his due.

Johann Bolyai published nothing more, though he continued his investigations. Notes found among his papers show that he was interested in the further extension of his ideas into space of three dimensions and also in the comparison of his Non-Euclidean Geometry with Spherical Trigonometry. It was this latter comparison which led him to the conviction that the Fifth Postulate could not be proved.[14] He was never thoroughly convinced, however, that investigations into space of three dimensions might not lead to the discovery of inconsistencies in the new geometry.

In 1848 Bolyai learned that the honor for the discovery of Non-Euclidean Geometry must be shared with still another. In that year he received information of Lobachewsky's discoveries and examined them critically. There was aroused in him the spirit of rivalry, and in an attempt to outshine Lobachewsky he began to labor in earnest again on what was to be his great work, the *Raumlehre*, which he had planned when he was publishing the *Appendix*. But this work was never completed.

31. Lobachewsky.

Although it was not until 1848 that Bolyai learned of the work of Nikolai Ivanovich Lobachewsky (1793–1856), the latter had discovered the new geometry and had actually published his conclusions as early as 1829, two or three years before the appearance in print of the *Appendix*. But there is ample evidence that he made his discoveries later than Bolyai made his.

Lobachewsky[15] took his degree at the University of Kasan in 1813.

[14] Stäckel, *loc. cit.*, Vol. I, p. 121.
[15] Perhaps the best account of Lobachewsky and his work is to be found in Friedrich Engel: *N. I. Lobatschefskij*, (Leipzig, 1898).

He was retained as instructor and later was promoted to a professorship. As a student there he had studied under Johann M. C. Bartels, who had been one of the first to recognize the genius of Gauss. Although Gauss and Bartels were close friends, there is no evidence that the latter, when he went to Kasan in 1807, carried with him and imparted to Lobachewsky any advanced views on the problem of parallels. Indeed, we know that Gauss himself at that early date was still working along conventional lines. The later discoveries of Lobachewsky seem to have been the results of his own initiative, insight and ability.

At any rate, along with the others, he was trying to prove the Fifth Postulate as early — or as late — as 1815. A copy of the lecture notes, taken by one of his students during that year and the two following, reveals only attempts to verify the Euclidean theory. It was not until after 1823 that he began to change his viewpoint, by which date, it will be recalled, Johann Bolyai had reached pretty well organized ideas about his new geometry.

In 1823 Lobachewsky had completed the manuscript for a textbook on elementary geometry, a text which was never published. This manuscript is extant. In it he made the significant statement that no rigorous proof of the Parallel Postulate had ever been discovered and that those proofs which had been suggested were merely explanations and were not mathematical proofs in the true sense. Evidently he had begun to realize that the difficulties encountered in the attempts to prove the Postulate arose through causes quite different from those to which they had previously always been ascribed.

The next three years saw the evolution of his new theory of parallels. It is known that in 1826 he read a paper before the physics and mathematics section of the University of Kasan and on that occasion suggested a new geometry in which more than one straight line can be drawn through a point parallel to a given line and the sum of the angles of a triangle is less than two right angles. Unfortunately the lecture was never printed and the manuscript has not been found.

But in 1829–30 he published a memoir on the principles of geometry in the *Kasan Bulletin*, referring to the lecture mentioned above, and explaining in full his doctrine of parallels. This memoir, the first account of Non-Euclidean Geometry to appear in print, attracted little attention in his own country, and, because it was printed in Russian, practically none at all outside.

Confident of the merit of his discoveries, Lobachewsky wrote a number of papers, more or less extensive, on the new theory of parallels, hoping thus to bring it to the attention of mathematicians all over the world. Perhaps the most important of these later publications was a little book entitled *Geometrische Untersuchungen zur Theorie der Parallellinien*,[16] written in German with the idea that it might for that reason be more widely read. A year before his death, although he had become blind, he wrote a complete account of his researches which was published in French under the title: *Pangéométrie ou précis de géométrie fondée sur une théorie générale et rigoureuse des parallèles*.[17] But he did not live to see his work accorded any wide recognition.

So slowly was information of new discoveries circulated in those days that Gauss himself did not learn of the advances made by Lobachewsky for a number of years, perhaps not until after the publication of the *Untersuchungen*. At any rate, it appears that by 1841 he knew of Lobachewsky and his work and was deeply impressed. In 1846 he wrote to Schumacher as follows:[18]

"I have recently had occasion to look through again that little volume by Lobatschefski (Geometrische Untersuchungen zur Theorie der Parallellinien, Berlin 1840, bei B. Funcke, 4 Bogen stark). It contains the elements of that geometry which must hold, and can with strict consistency hold, if the Euclidean is not true. A certain Schweikardt[19] calls such geometry Astral Geometry, Lobatschefsky calls it Imaginary Geometry. You know that for fifty-four years now (since 1792) I have held the same conviction (with a certain later extension, which I will not mention here). I have found in Lobatschefsky's work nothing that is new to me, but the development is made in a way different from that which I have followed, and certainly by Lobatschefsky in a skillful way and in truly geo-

[16] See Lobatschewsky: *Geometrical Researches on the Theory of Parallels*, translated by G. B. Halsted (Austin, Texas, 1891).

[17] See David Eugene Smith: *A Source Book in Mathematics*, p. 360 (New York, 1929).

[18] Engel and Stäckel: *Die Theorie der Parallellinien von Euklid bis auf Gauss*, p. 235.

[19] See Section 32.

metrical spirit. I feel that I must call your attention to the book, which will quite certainly afford you the keenest pleasure."

By 1848 Wolfgang Bolyai had heard in some way of Lobachewsky's investigations. In January of that year he wrote to Gauss, asking for the name of the book by the Russian mathematician. Gauss recommended "that admirable little work," the *Geometrische Untersuchungen*, as containing an adequate exposition of the theory and as being easily obtainable. Thus Wolfgang and, through him, Johann became acquainted with the geometry of Lobachewsky.

That Johann received this information about the work of the Russian geometer philosophically enough is evinced by remarks found in his unpublished notes entitled: *Bemerkungen über Nicolaus Lobatchefskij's Geometrische Untersuchungen*. He wrote in part:[20]

"Even if in this remarkable work different methods are followed at times, nevertheless, the spirit and result are so much like those of the *Appendix* to the *Tentamen matheseos* which appeared in the year 1832 in Maros-Vásárhely, that one cannot recognize it without wonder. If Gauss was, as he says, surprised to the extreme, first by the *Appendix* and later by the striking agreement of the Hungarian and Russian mathematicians: truly, none the less so am I.

"The nature of real truth of course cannot but be one and the same in Maros-Vásárhely as in Kamschatka and on the Moon, or, to be brief, anywhere in the world; and what one finite, sensible being discovers, can also not impossibly be discovered by another."

But, regardless of these reflections, for a time at least, Bolyai entertained the suspicion that somehow Lobachewsky had learned of his own discoveries, possibly through Gauss, and had then, after some revision, published them. His attitude later, however, became somewhat more lenient. As a matter of fact, there seems to be no evidence that Lobachewsky ever heard of Bolyai.

32. Wachter, Schweikart and Taurinus.

No satisfactory record, however brief, of the discovery of Non-Euclidean Geometry will fail to include the names of Wachter, Schweikart and Taurinus. We insert here short accounts of their contributions, before turning our attention to further developments due to Riemann and others.

[20] Stäckel: *Wolfgang und Johann Bolyai, Geometriche Untersuchungen*, Vol. I, p. 140 (Leipzig and Berlin, 1913).

Friedrich Ludwig Wachter (1792-1817), Professor of Mathematics in the Gymnasium at Dantzig, studied under Gauss at Göttingen in 1809. His attempts to prove the Fifth Postulate led to the publication in 1817 of a paper[21] in which he attempted to prove that through any four points in space, not lying in one plane, a sphere can be constructed. This plan of investigation was obviously suggested by the fact that the Postulate can be proved once it is established that a circle can be drawn through any three non-collinear points. Although his arguments were unsound, some of his intuitive deductions in this paper, and in a letter[22] written to Gauss in 1816, are worthy of recognition. Among other things, he remarked that, even if Euclid's Postulate is denied, spherical geometry will become Euclidean if the radius of the sphere is allowed to become infinite, although the limiting surface *is not a plane*. This was confirmed later by both Bolyai and Lobachewsky.

Wachter lived only twenty-five years. His brief investigations held much promise and exhibited keen insight. Had he lived a few years longer he might have become the discoverer of Non-Euclidean Geometry. As it was, his influence was probably considerable. Just at the time when he and Gauss were discussing what they called *Anti-Euclidean* Geometry, the latter began to show signs of a change of viewpoint. In 1817, writing to H. W. M. Olbers, his associate and a noted astronomer, Gauss was led to remark, after mentioning Wachter, and commending his work despite its imperfections,[23] "I keep coming closer to the conviction that the necessary truth of our geometry cannot be proved, at least *by* the human intellect *for* the human intellect. Perhaps in another life we shall arrive at other insights into the nature of space which at present we cannot reach. Until then we must place geometry on an equal basis, not with arithmetic, which has a purely *a priori* foundation, but with mechanics."

It will be recalled[24] that Gauss, in a letter to Schumacher, mentioned "a certain Schweikardt." The one referred to was Ferdinand

[21] For this paper and some of Wachter's letters see Stäckel: *Friedrich Ludwig Wachter, ein Beitrag zur Geschichte der Nichteuklidischen Geometrie*, Mathematische Annalen, Vol. 54, pp. 49-85 (1901).
[22] Stäckel, *ibid.*, p. 61.
[23] Stäckel, *ibid.*, p. 55.
[24] See Section 31.

Karl Schweikart (1780–1859), who from 1796 to 1798 was a student
of law at Marburg. As he was keenly interested in mathematics,
he took advantage of the opportunity while at the university to
listen to the lectures of J. K. F. Hauff, who was somewhat of an
authority on the theory of parallels. Schweikart's interest in this
theory developed to such an extent that in 1807 there appeared his
only published work of a mathematical nature, *Die Theorie der
Parallellinien nebst dem Vorschlage ihrer Verbannung aus der Geometrie.*[25]
In spite of its title, this book offered nothing particularly novel
and was written along quite conventional lines. In it he mentioned
both Saccheri and Lambert. His acquaintance with the work of
these men doubtless affected the character of his later investigations.
In 1812 Schweikart went to Charkow; the year 1816 found him in
Marburg again, where he remained until 1820 when he became Professor of Jurisprudence at Königsberg.

In 1818 he handed to his friend Gerling, student of Gauss and
Professor of Astronomy at Marburg, a brief outline of his ideas
about a new geometry in which the Parallel Postulate was denied,
and asked him to forward it to Gauss for his criticism. In this
memorandum he asserted that there are two kinds of geometry,
Euclidean and *Astral,* and that in the latter the sum of the angles
of a triangle is less than two right angles; the smaller the angle-sum,
the greater the area of the triangle; that the altitude of an isosceles
right triangle increases as the sides increase, but can never become
greater than a certain length called the *Constant;* that, when this
Constant is taken as infinite, Euclidean Geometry results. This
outline was probably the first explicit description of a Non-Euclidean
Geometry, regarded as such. The ideas came to Schweikart before
1816, while he was still in Charkow. At that early date both
Bolyai and Lobachewsky were still carrying on their investigations
from the traditional viewpoint.

In his reply to Gerling, Gauss commended Schweikart highly.
"The memorandum of Professor Schweikardt has brought me the
greatest pleasure," he wrote, "and in regard to it please extend to
him my sincerest compliments. It might almost have been written
by myself."

Schweikart did not publish the results of any of his investigations.

[25] See Engel and Stäckel, *loc. cit.*, p. 243.

But he did encourage his sister's son, Franz Adolph Taurinus (1794–1874), to take up the study of parallels, suggesting that he give some thought to the Astral Geometry which Gauss had praised so highly. Taurinus, after studying jurisprudence for a brief time, had settled down in Köln to spend a long life of leisure, with ample time to devote to varied intellectual interests. In 1824, when he first began a systematic investigation of the problem of parallels, he found himself not in accord with his uncle's ideas. That he hoped, at this early point in his researches, to be able to prove the Fifth Postulate is nothing out of the ordinary. The remarkable fact is that, although as a consequence of his independent investigations he was one of the first to obtain a view of Non-Euclidean Geometry, nevertheless throughout his life he continued to believe that the Euclidean Hypothesis was the only one of the three which would lead to a valid geometry.

In 1825, soon after he had received from Gauss the complimentary and encouraging letter which was translated in full in Section 29, appeared his first book, *Theorie der Parallellinien*.[26] Here he attacked the problem from the Non-Euclidean viewpoint, rejecting the Hypothesis of the Obtuse Angle and, using the Hypothesis of the Acute Angle, encountered the Constant of Schweikart. These investigations led him to ideas which were not in accord with his concept of space and he was impelled to reject the latter hypothesis also, although he appeared to recognize the consequences as logically sound.

Shortly after the publication of his first book he learned that Saccheri and Lambert had both preceded him along the route he had followed. So he produced another book in 1826, his *Geometriae Prima Elementa*,[27] in which he modified his method of attack. It was in the appendix of this work that he made his most important contributions. Here he developed many of the basic formulas for Non-Euclidean Trigonometry. In the familiar formulas of spherical trigonometry he replaced the real radius of the sphere by an imaginary one. The modified formulas, remarkably enough, describe the geometry which arises under the Hypothesis of the Acute Angle. Lambert had previously investigated trigonometric functions with

---

[26] For an excerpt, see Engel and Stäckel, *loc. cit.*, p. 255.
[27] Engel and Stäckel, *loc. cit.*, p. 267.

imaginary arguments, and in that connection had developed to some extent the theory of hyperbolic functions, but there is no evidence that he tried to use these ideas in his study of parallels. It will be recalled that he had surmised that this geometry might be verified on a sphere of imaginary radius.[28] Taurinus did not use hyperbolic functions; instead he exhibited the real character of his formulas through the medium of exponents and logarithms. Consequently he called the geometry *Logarithmisch-Sphärischen Geometrie*. He, as had Lambert, recognized the correspondence between spherical geometry and that which arises if the Hypothesis of the Obtuse Angle is used. In addition, he noted that his Logarithmic-Spherical Geometry became Euclidean when the radius of the sphere was made infinite.

Although his reluctance to recognize this geometry as valid on a plane persisted, Taurinus seemed to be fully aware of the importance of his discoveries, from the theoretical viewpoint, in the study of parallels. His *Geometriae Prima Elementa* received little recognition. In his disappointment, he burned the remaining copies.

33. Riemann.

Neither Bolyai nor Lobachewsky lived to see his work accorded the recognition which it merited. This delay can be attributed to several factors: the slow passage of ideas from one part of the world to another, the language barriers, the Kantian space philosophy, the two-thousand-year dominance of Euclid, and the relative obscurity of the discoverers of Non-Euclidean Geometry. The new geometry attracted little attention for over thirty-five years until, in 1867, Richard Baltzer, in the second edition of his *Elemente der Mathematik*, inserted a reference to it and its discoverers, and also persuaded Hoüel to translate their writings into French.

But in the meantime a new figure had appeared. Born at about the time of the discovery of Non-Euclidean Geometry, George Friedrich Bernhard Riemann (1826-1866) grew to young manhood with the intention of studying theology. But when he entered Göttingen for that purpose he discovered that mathematics was his forte and gave up theology. He studied under Gauss and became the outstanding student in the long teaching career of that great mathematician. Later he went to Berlin to study with Dirichlet, Jacobi,

[28] See Section 23.

Steiner and others, but returned to Göttingen in 1850 to study physics and take his degree there the following year.

We have already quoted[29] from the remarkable probationary lecture, *Über die Hypothesen welche der Geometrie zu Grunde liegen*, which he delivered in 1854 before the Philosophical Faculty at Göttingen, and in which he pointed out that space need not be infinite, though regarded as unbounded. Thus he suggested indirectly a geometry in which no two lines are parallel and the sum of the angles of a triangle is greater than two right angles. It will be recalled that, in the rejection of the Hypothesis of the Obtuse Angle by earlier investigators, the infinitude of the line had been assumed.

But Riemann, in this memorable dissertation, did more than that; he called attention to the true nature and significance of geometry and did much to free mathematics of the handicaps of tradition. Among other things, he said,[30] "I have in the first place .... set myself the task of constructing the notion of a multiply extended magnitude out of general notions of magnitude. It will follow from this that a multiply extended magnitude is capable of different measure-relations, and consequently that space is only a particular case of a triply extended magnitude. But hence flows as a necessary consequence that the propositions of geometry cannot be derived from general notions of magnitude, but that the properties which distinguish space from other triply extended magnitudes are only to be deduced from experience. Thus arises the problem, to discover the simplest matters of fact from which the measure-relations of space may be determined; a problem which from the nature of the case is not completely determinate, since there may be several systems of matters of fact which suffice to determine the measure-relations of space — the most important system for our present purpose being that which Euclid has laid down as a foundation. These matters of fact are — like all matters of fact — not necessary, but only of empirical certainty; they are hypotheses. We may therefore investigate their probability, which within the limits of observation is of course very great, and inquire about the justice of their extension beyond the limits of observation, on the side both of the infinitely great and of the infinitely small."

29 See Section 6.
30 Clifford's translation, *Nature*, Vol. VIII, 1873. See also David Eugene Smith, *A Source Book in Mathematics*, pp. 411–425 (New York, 1929).

Stressing the importance of the study of the properties of things from the infinitesimal standpoint, he continued, "The questions about the infinitely great are for the interpretation of nature useless questions. But this is not the case with the questions about the infinitely small. It is upon the exactness with which we follow phenomena into the infinitely small that our knowledge of their causal relations essentially depends. The progress of recent centuries in the knowledge of mechanics depends almost entirely on the exactness of the construction which has become possible through the invention of the infinitesimal calculus, and through the simple principles discovered by Archimedes, Galileo and Newton, and used by modern physics. But in the natural sciences which are still in want of simple principles for such constructions, we seek to discover the causal relations by following the phenomena into great minuteness, so far as the microscope permits. Questions about the measure-relations of space in the infinitely small are not therefore superfluous questions."

Thus began a second period in the development of Non-Euclidean Geometry, a period characterized by investigations from the viewpoint of differential geometry in contrast with the synthetic methods previously used. Riemann's memoir dealt almost altogether with generalities and was suggestive in nature. The detailed investigations along these lines were carried out by others, notably Helmholtz, Lie and Beltrami. The contributions of the physicist, Helmholtz, remarkable as they were, required for rigor the finishing touches of a mathematician. These thorough investigations were made by Lie, using the idea of groups of transformations. To Beltrami goes the credit of offering the first proof of the consistency of Non-Euclidean Geometry. Although Bolyai and Lobachewsky had encountered no contradiction in their geometry as far as their investigations had gone, there still remained the possibility that some such inconsistency might arise as the research continued. Beltrami showed how this geometry can be represented, with restrictions, on a Euclidean surface of constant curvature, and thus how any inconsistency discovered in the geometry of Bolyai and Lobachewsky will lead to a corresponding one in Euclidean Geometry.

34. Further Developments.

The work of this second period was excellent and the results were far reaching and significant, but it remained for a third period — one with which are associated the names of Cayley, Klein and Clifford — to supply what was still needed in the way of the unification and interpretation of the Non-Euclidean Geometries. The beautiful classification of these geometries from the projective-metric viewpoint and the recognition of the roles which they play in the rounding out of a logical category led to their complete justification and thus brought to a triumphant close the long struggle with the Fifth Postulate.

In his notable *Sixth Memoir upon Quantics*,[31] Cayley, in 1859, showed how the notion of distance can be established upon purely descriptive principles. These ideas were developed and interpreted from the standpoint of Non-Euclidean Geometry by Felix Klein in two monographs[32] appearing in 1871 and 1873. It was he who suggested calling the geometries of Bolyai and Lobachewsky, Riemann, and Euclid, respectively, *Hyperbolic*, *Elliptic* and *Parabolic*, a terminology almost universally accepted and which we shall use from this point on. The names were suggested by the fact that a straight line contains two infinitely distant points under the Hypothesis of the Acute Angle, none under the Hypothesis of the Obtuse Angle, and only one under the Hypothesis of the Right Angle.

More recently investigators have confined their attentions largely to careful scrutiny of the foundations of geometry and to the precise formulation of the sets of axioms. Following the lead of Pasch, such men as Hilbert, Peano, Pieri, Russell, Whitehead and Veblen have gone far in placing geometry, both Euclidean and Non-Euclidean, as well as mathematics in general, on a firm logical basis.

35. Conclusion.

In the following pages we shall take up first a study of Synthetic Hyperbolic Geometry. This will be followed by an investigation of the trigonometry of the Hyperbolic Plane and that, in turn, by a

---

[31] See Cayley's *Collected Mathematical Papers*, Vol. II, pp. 561–592 (Cambridge, 1889).
[32] See his *Gesammelte Mathematische Abhandlungen*, Vol. I, pp. 254–305 and pp. 311–343 (Berlin, 1921).

brief treatment from the viewpoint of analytic geometry and calculus.

Our examination of Elliptic Geometry will be less extensive. Its
development, like that of most of the later work in Non-Euclidean
Geometry, depends upon the use of concepts more advanced than
those which we wish to draw upon here.

In passing, we remark that there are two types of Elliptic Geometry. The one suggested in Section 6 is probably the one which
Reimann had in mind. Geometry on this Elliptic Plane has an
exact analogue in the geometry on a sphere, the great circles being
regarded as straight lines. The other type, in many respects the
more interesting and important, was suggested later by Klein.
In this geometry, two points always determine a straight line, and
in other respects it more nearly resembles Euclidean Geometry.

IV

HYPERBOLIC PLANE GEOMETRY

"It is quite simple, ... , but the way things come out of one another is quite lovely." — CLIFFORD

36. Introduction.

There is much to be said in favor of carefully formulating, at this point, a set of explicitly stated assumptions as a foundation for the study of the geometry of the Hyperbolic Plane.[1] For very mature students and for those who have been over the ground before, this is doubtless the best procedure. But for others, such a precise, rigorously logical treatment would only prove confusing. For this reason, keeping in mind our objectives, we propose to follow the path of the pioneers, to avail ourselves of the familiar foundations of Euclidean Geometry, replacing the Fifth Postulate by a contradictory one, and making such other changes as may consequently be forced upon us. Thus all of the Euclidean propositions which do not depend upon the Fifth Postulate, in particular the first twenty-eight, are immediately available. Nor are simplicity and economy the only advantages of such an approach. This presentation of the material, in the way it was acquired, has, we believe, the sanction of sound pedagogical and psychological principles. Refinement and extreme rigor may very well come later.

---

[1] The student who wishes to examine Hyperbolic Geometry from this more critical viewpoint is referred to Hilbert, *Grundlagen der Geometrie*, 7th edition, p. 159 (Leipzig and Berlin, 1930).

37. The Characteristic Postulate of Hyperbolic Geometry and Its Immediate Consequences.

In Euclidean Plane Geometry, the Fifth Postulate is essentially equivalent to the statement that through a given point, not on a given line, one and only one line can be drawn which does not intersect the given line. In its place we introduce the following as the *Characteristic Postulate* of Hyperbolic Plane Geometry.

POSTULATE. *Through a given point, not on a given line, more than one line can be drawn not intersecting the given line.*

We have already observed that, if there is more than one line through the given point not intersecting the given line, there is an infinite number of them. If P (Fig. 21) is the given point, l the given line and AB and CD two lines through P which do not intersect l, then no line, such as EF, lying within the vertical angles APC and DPB, which do not contain the perpendicular PQ from

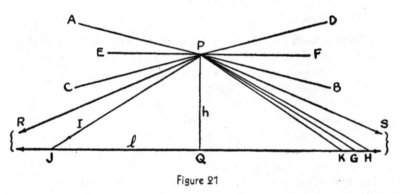

Figure 21

P to l, will cut l. For if EF, produced for example to the right, were to cut l, then, since PF and the perpendicular PQ would both intersect l, PB would also have to intersect l by the Axiom of Pasch.

If one starts with the perpendicular PQ from P to l and allows PQ to rotate about P in either direction, say counterclockwise, it will continue to intersect l awhile and then cease to intersect it. Thus a situation is reached in which the lines through P are divided into two sets, those which cut l and those which do not, each line of the first set preceding each line of the second. Under these circumstances, the Postulate of Dedekind asserts that there exists a line through *P* which brings about this division of the lines into the two sets. Since this line itself either cuts *l* or does not cut it, it must either be the last of the lines which intersect *l* or the first of those which do not. But there is no last cutting line. For, if one assumes that *PG* is the last of the cutting lines, and measures off any distance *GH* on the side of *G* opposite *Q*, then *PH* is a cutting line and a contradiction has been reached. Hence the dividing line is the first of those which do not intersect *l*. A similar situation is encountered if *PQ* is rotated clockwise. Thus there are two lines *PR* and *PS*, through *P*, which do not cut *l*, and which are such that every line through *P*, lying within the angle *RPS*, does cut *l*.

Furthermore, the angles *RPQ* and *SPQ* are equal. If they are not, one of them is the greater, say *RPQ*. Measure angle *IPQ* equal to angle *SPQ*. Then *PI* will cut *l* in a point *J*. Measure off on *l*, on the side of *Q* opposite *J*, *QK* equal to *QJ*. Draw *PK*. It follows from congruent right triangles that angle *QPK* equals angle *QPJ* and hence angle *QPS*. But *PS* does not intersect *l* and a contradiction has been reached. One concludes that angles *RPQ* and *SPQ* are equal.

It can easily be shown that these two angles are acute. If they were right angles, *PR* and *PS* would lie on the same straight line and this line would be the line through *P* perpendicular to *PQ*. But that perpendicular does not intersect *l* (Euclid I, 28), and furthermore it is not the only line through *P* which does not intersect *l*. Consequently there will be lines through *P* within the angle *RPS* which do not cut *l* under these circumstances and again a contradiction is encountered.

These results can be summarized in the following theorem:

Theorem. If *l* is any line and *P* is any point not on *l*, then there are always two lines through *P* which do not intersect *l*, which make equal acute angles with the perpendicular from *P* to *l*, and which are such that every line through *P* lying within the angle containing that perpendicular intersects *l*, while every other line through *P* does not.

All of the lines through P which do not meet l are parallel to l from the viewpoint of Euclid. Here, however, we wish to recognize the peculiar character of the two described in the theorem above. These two lines we call the two *parallels* to l through P, and designate the others as *non-intersecting* with reference to l. We shall discover shortly that the size of the angle which each of the two parallels makes with the perpendicular from P to l depends upon the length h of this perpendicular. The angle is called the *angle of parallelism* for the distance h and will be denoted by $\Pi(h)$ in order to emphasize the functional relationship between the angle and the distance. On occasion it will be found possible and convenient to distinguish between the two parallels by describing one as the *right-hand*, the other as the *left-hand*, parallel.

EXERCISE

If two lines BA and BC are both parallel to line l, show that the bisector of angle ABC is perpendicular to l.

38. Elementary Properties of Parallels.

Certain properties of Euclidean parallels hold also for parallels in Hyperbolic Geometry. Three of these are described in the following theorems:

Theorem 1. If a straight line is the parallel through a given point in a given sense to a given line, it is, at each of its points, the parallel in the given sense to the given line.

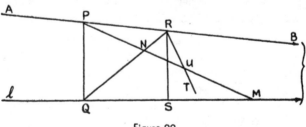

Figure 22

If *AB* (Fig. 22) is one of the two parallels, say the right-hand parallel, to line *l*, through *P*, we wish to prove that it is the right-hand parallel to *l* through any of its points. There are two cases to be considered.

CASE I. Let *R* be any point on *AB* on the side of *P* in the direction of parallelism. Draw *PQ* and *RS* perpendicular to *l*. We have to show that every line through *R* passing within the angle *SRB* intersects *l*. Let *RT* be any such line and select on it any point *U*. Draw *PU* and *RQ*. *PU* must cut *l* in a point *M* and, by Pasch's axiom, must cut *QR* in a point *N*. Again resorting to the Axiom of Pasch, we find that *RU* intersects *QM*, since it intersects segment *NM*, but not segment *QN*.

CASE II. Let *R* be any point on *AB* on the side of *P* in the direction opposite that of parallelism. In this case, all that is needed is to choose *U* as any point on *TR* produced through *R*, use the same lettering, and follow the same plan. The details are left to the reader.

Theorem 2. If one line is parallel to a second, then the second is parallel to the first.

Let *AB* (Fig. 23) be the right-hand parallel to *CD* through *P*, draw *PQ* perpendicular to *CD* and *QR* perpendicular to *AB*. The point *R* will fall to the right of *P* (Euclid I, 16). Then, in order to prove that *CD*, which does not cut *AB*, is parallel to *AB*, we have to show that every line through *Q* and lying within the angle *RQD* intersects *RB*.

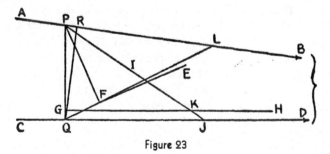

Figure 23

Let QE be any such line and construct PF perpendicular to QE. F will lie on the side of Q on which E lies. On PQ measure PG equal to PF. The point G will lie between P and Q since PF is shorter than PQ. Draw GH perpendicular to PQ at G. Next construct angle GPI equal to angle FPB and produce PI until it intersects CD at J. Since GH cuts side PQ of triangle PQJ, but not side QJ, it cuts PJ at some point K. On PB measure PL equal to PK and join F and L. Since triangles PGK and PFL are congruent, angle PFL is a right angle. But PFE is a right angle. Hence QE intersects RB at L.

Theorem 3. If two lines are both parallel to a third line in the same direction, then they are parallel to one another.

First consider the case in which the third line lies between[2] the other two. Let AB and CD (Fig. 24) both be parallel to EF in the same direction and let AC cut EF at G. Draw any line AH through A and passing within the angle CAB. This line will cut EF in a

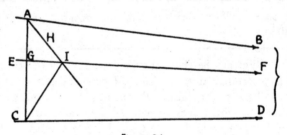

Figure 24

point I. Draw CI. Since EF is parallel to CD, AI produced will intersect CD. Since AB does not cut CD, but every line through A lying within angle CAB does, it follows that AB and CD are parallel.

Next consider the case in which the two lines are on the same side of the third. Let AB and CD (Fig. 25) both be parallel to EF in the same direction. Assume that AB is not parallel to CD in the designated direction. Then through any point G of AB draw the parallel GH to CD in that direction. It follows from the first case that GH

---

[2] See Section 9.

is parallel to *EF*. But only one parallel in that direction can be drawn to *EF* through *G*. Therefore *GH* must coincide with *AB*, and *AB* is parallel to *CD*.

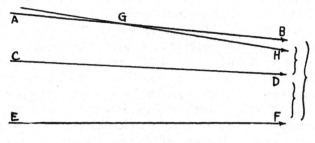

Figure 25

39. Ideal Points.

We wish to introduce at this point an important concept in connection with parallel lines. Two intersecting lines have a point in common, but two parallel lines do not, since they do not intersect. However, two parallel lines do have *something* in common. It is convenient to recognize this relationship by saying that two parallel lines have in common, or intersect in, an *ideal point*.[3] Thus all of the lines parallel in the same sense to any line, and consequently parallel to one another, will be thought of as being concurrent in an ideal point and constituting a sheaf of lines with an ideal vertex. Every line contains thus, in addition to its *ordinary* or *actual* points, two ideal points through which pass all of the lines parallel to it in the two directions.

While ideal points are concepts, so also, for that matter, are ordinary points. The introduction of these ideal elements is primarily a matter of convenient terminology. To say that two lines intersect in an ideal point is merely another way of saying that they are parallel; to refer to the line joining an ordinary point to one of the ideal points of a certain line amounts to referring to the line through the ordinary point parallel to that line in the sense designated. But we shall not be surprised to find these new entities

---

[3] Also called, more frequently than not, a *point at infinity* or *infinitely distant point*.

assuming more and more significance as we go on. In the history of mathematics can be found more than one example of an idea introduced for convenience developing into a fundamental concept. As a matter of fact, the use of such ideal elements has been an important factor in the development of geometry and in the interpretation of space. We shall return to this later.

It will gradually be recognized that, in so far as we are concerned with purely descriptive properties, we need not discriminate between ordinary and ideal points. Two distinct points, for example, determine a line, regardless of whether both points are ordinary, both ideal, or one ordinary and the other ideal. In no case is this more strikingly illustrated than in that of the triangle with two vertices ordinary points and the third ideal. We study this figure next.

40. Some Properties of an Important Figure.

The figure formed by two parallel lines and the segment joining a point of one to a point of the other plays an important role in what is to follow. Let $A\Omega$ and $B\Omega$ (Fig. 26) be any two parallel lines. Here we follow the convention of using the large Greek letters (generally Ω) to designate ideal points. Let A, any point of the first, and B, any point of the second, be joined. The resulting figure is in the nature of a triangle with one of its vertices an ideal point; it has many properties in common with ordinary triangles. We prove first that Pasch's Axiom holds for such a triangular figure.

Theorem 1. If a line passes within the figure $AB\Omega$ through one of the vertices, it will intersect the opposite side.

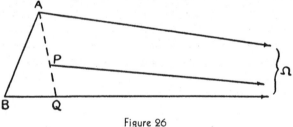

Figure 26

Let P (Fig. 26) be any point within the figure. Then AP and BP will intersect $B\Omega$ and $A\Omega$, respectively, as a consequence of the parallelism of $A\Omega$ and $B\Omega$. Let AP cut $B\Omega$ in point Q. Draw $P\Omega$. This line, if produced, will intersect segment AB by Pasch's Axiom.

Theorem 2. If a straight line intersects one of the sides of $AB\Omega$, but does not pass through a vertex, it will intersect one and only one of the other two sides.

If the line intersects $A\Omega$ or $B\Omega$, the theorem is easily proved. If it intersects AB in a point R, one has but to draw $R\Omega$ and make use of Theorem 1. The details are left to the reader.

The familiar exterior angle theorem also holds for these figures.

Theorem 3. The exterior angles of $AB\Omega$ at A and B, made by producing AB, are greater than their respective opposite interior angles.

Let AB (Fig. 27) be produced through B to C. It is to be proved that angle $CB\Omega$ is greater than angle $BA\Omega$. Through B draw BD, making angle CBD equal to angle $BA\Omega$. BD cannot cut $A\Omega$, for in this case there would be formed a triangle with an exterior angle equal to one of the opposite interior angles. Furthermore, it cannot

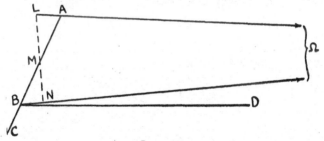

Figure 27

coincide with $B\Omega$ and thus be parallel to $A\Omega$. To show this, draw MN from M, the midpoint of AB, perpendicular to $B\Omega$. Measure off on $A\Omega$, on the side of AB opposite N, a segment AL equal to BN. Draw ML. It is easy to show that MN and ML lie in the same straight line since triangles MNB and MLA are congruent if BD coincides with $B\Omega$. In this case it follows that LN is perpendicular to both $A\Omega$ and $B\Omega$ and that the angle of parallelism for the distance LN is a right angle. But this is absurd. Therefore BD will lie within the angle $CB\Omega$ and this angle is greater than angle CBD. Hence angle $CB\Omega$ is greater than angle $BA\Omega$.

Next we describe the conditions under which two such figures, $AB\Omega$ and $A'B'\Omega'$, are congruent.

Theorem 4. If AB and $A'B'$ are equal, and angle $BA\Omega$ is equal to angle $B'A'\Omega'$, then angle $AB\Omega$ is equal to angle $A'B'\Omega'$ and the figures are congruent.

If angles $AB\Omega$ and $A'B'\Omega'$ (Fig. 28) are unequal under the given conditions, one of them is the greater, say $AB\Omega$. Construct angle ABC equal to angle $A'B'\Omega'$. Let BC cut $A\Omega$ in point D. Measure $A'D'$ on $A'\Omega'$ equal to AD and draw $B'D'$. Then triangles ABD and $A'B'D'$ are congruent. Thus angle $A'B'D'$ is equal to angle ABD and hence to angle $A'B'\Omega'$. The resulting contradiction leads us to the conclusion that angles $AB\Omega$ and $A'B'\Omega'$ are equal.

It is perhaps superfluous to remark that this theorem still holds if one of the figures is reversed, either by drawing the parallels in the opposite direction or by interchanging the two angles.

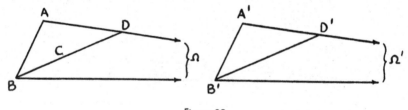

Figure 28

Theorem 5. If angles $BA\Omega$ and $B'A'\Omega'$ are equal and also angles $AB\Omega$ and $A'B'\Omega'$, then segments AB and $A'B'$ are equal and the figures are congruent.

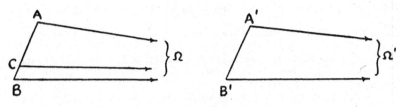

Figure 29

If, under the given conditions, AB and $A'B'$ (Fig. 29) are not equal, one of them, say AB, is the greater. Measure off on AB the segment AC equal to $A'B'$. Draw $C\Omega$. Then $AC\Omega$ and $A'B'\Omega'$ are congruent and angles $AC\Omega$ and $A'B'\Omega'$ are equal. It follows that angle $AC\Omega$ is equal to angle $AB\Omega$. But this contradicts Theorem 3. Therefore AB and $A'B'$ are equal.

Theorem 6. If segments AB and $A'B'$, angles $AB\Omega$ and $BA\Omega$, and angles $A'B'\Omega'$ and $B'A'\Omega'$, are equal, then all four angles are equal to one another and the figures are congruent.

Assume that the four angles are not equal to one another. Then one pair of equal angles, say $AB\Omega$ and $BA\Omega$, will be greater than the other pair. Construct angles ABC and BAD (Fig. 30) equal to angle $B'A'\Omega'$. BC and AD will intersect in a point E. Measure off on

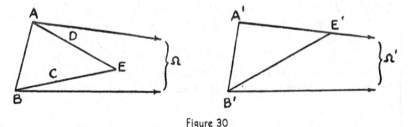

Figure 30

$A'\Omega'$ segment $A'E'$ equal to AE and draw $B'E'$. Then triangles ABE and $A'B'E'$ are congruent. A contradiction is reached when we conclude that angles $A'B'E$ and $A'B'\Omega'$ are equal. Thus all four angles must be equal.

EXERCISES

1. In the figure $AB\Omega$, the sum of angles $AB\Omega$ and $BA\Omega$ is always less than two right angles.

2. If a transversal meets two lines, making the sum of the interior angles on the same side equal to two right angles, then the two lines cannot meet and are not parallel; they are non-intersecting lines.

3. Given two parallel lines, $A\Omega$ and $B\Omega$, and two other lines, $A'C'$ and $B'D'$, prove that if segments AB and $A'B'$, angles $BA\Omega$ and $B'A'C'$, and angles $AB\Omega$ and $A'B'D'$, are equal, then $A'C'$ and $B'D'$ are parallel.

4. If angles $AB\Omega$ and $BA\Omega$ are equal, the figure is in the nature of an isosceles triangle with vertex an ideal point. Prove that, if M is the midpoint of AB, $M\Omega$ is perpendicular to AB. Show also that the perpendicular to AB at M is parallel to $A\Omega$ and $B\Omega$ and that all points on it are equally distant from those two lines.

5. Prove that if, in figure $AB\Omega$, the perpendicular to AB at its midpoint is parallel to $A\Omega$ and $B\Omega$, then the angles at A and B are equal.

6. If, for two figures $AB\Omega$ and $A'B'\Omega'$, angles $AB\Omega$ and $A'B'\Omega'$ are equal but segment AB is greater than segment $A'B'$, then angle $BA\Omega$ is smaller than angle $B'A'\Omega'$.

41. The Angle of Parallelism.

From Theorem 4 of the preceding section, it is evident that the angle of parallelism $\Pi(h)$ for any given distance h is constant. Furthermore, as a consequence of Theorem 3, it follows that

$$h_1 \gtrless h_2$$

implies that

$$\Pi(h_1) \lessgtr \Pi(h_2).$$

We know, from the theorem of Section 37, that every distance has a corresponding angle of parallelism. It has just been pointed out that this angle is always the same for any given distance, that the angle increases as the distance decreases and decreases as the distance increases. Presently it will be shown that to every acute angle there corresponds a distance for which the angle is the angle of parallelism. In this case equal angles must, of course, have equal corresponding distances. Putting these results together, we conclude that

$$\lim_{\delta \to 0} [\Pi(h + \delta) - \Pi(h)] = 0,$$

and consequently that $\Pi(h)$ varies continuously if h does.

It should perhaps be noted here that, so far, no particular unit has been specified for measuring either distances or angles. The functional relationship implied has been purely geometric. Later, when definite units have been adopted, the analytic form of $\Pi(h)$

will be obtained. However, as h approaches zero, $\Pi(h)$ approaches a right angle, and we may write

$$\Pi(0) = \frac{\pi}{2},$$

where π is, for the present, used merely as a symbol to denote a straight angle. As h becomes infinite, $\Pi(h)$ approaches zero, or, in the conventional notation,

$$\Pi(\infty) = 0.$$

Moreover, there is no reason why we should not attach a meaning to $\Pi(h)$ for h negative. There is nothing which compels us to do this; we do it solely because it will prove convenient. Such a generalization will enable us to avoid certain exceptions later on. The definition of $\Pi(h)$, when h is negative, is a matter of choice, but we shall choose methodically.

As h, being positive, decreases, $\Pi(h)$ increases; when h is zero, $\Pi(h)$ is a right angle. If we think of h as continuing to decrease, becoming negative, we naturally choose to regard $\Pi(h)$ as continuing to increase and becoming obtuse. Briefly, $\Pi(-h)$ is defined by the relation

$$\Pi(h) + \Pi(-h) = \pi.$$

42. The Saccheri Quadrilateral.

It will be recalled that, as a basis for his investigations, Saccheri made systematic use of a quadrilateral formed by drawing equal perpendiculars at the ends of a line segment on the same side of it and connecting their extremities. This birectangular, isosceles quadrilateral is commonly called a *Saccheri Quadrilateral*. We shall study some of its properties. The side adjacent to the two right angles is known as the *base*, the opposite side as the *summit* and the angles adjacent to the summit as the *summit angles*.

Theorem. The line joining the midpoints of the base and summit of a Saccheri Quadrilateral is perpendicular to both of them; the summit angles are equal and acute.

Let *AB* (Fig. 31) be the base of the Saccheri Quadrilateral *ABCD*. Join *M* and *H*, the midpoints of base and summit, respectively, and construct *CM* and *DM*. It is not difficult to prove the congruence of

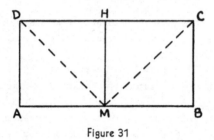

Figure 31

the triangles *DAM* and *CBM*, and then, in turn, of triangles *DHM* and *CHM*. The equality of the summit angles follows as well as that of the angles made by *MH* with *DC* and with *AB*.

Corollary. The base and summit of a Saccheri Quadrilateral are non-intersecting lines.

That the summit angles are acute, and consequently that Hyperbolic Geometry is the geometry of Saccheri's Hypothesis of the Acute Angle, is proved as follows:

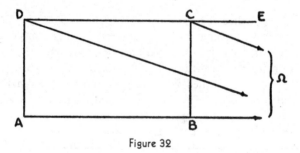

Figure 32

Let *D*Ω and *C*Ω be parallels to *AB*, in the same sense, through *D* and *C*, the extremities of the summit of the Saccheri Quadrilateral *ABCD* (Fig. 32). Let *E* be any point on *DC* produced through *C*.

Then the parallels $D\Omega$ and $C\Omega$ will lie within the angles ADC and BCE, respectively, since DC is non-intersecting with regard to AB. Angles $AD\Omega$ and $BC\Omega$ are equal, being angles of parallelism for equal distances. Furthermore, in the figure $CD\Omega$, the exterior angle $EC\Omega$ is greater than the opposite interior angle $CD\Omega$. Thus angle BCE is greater than angle ADC and hence greater than angle DCB. Consequently the equal summit angles are acute.

43. The Lambert Quadrilateral.

The reader will recollect that Lambert used, as a fundamental figure in his researches, a quadrilateral with three of its angles right angles. This trirectangular quadrilateral, which we shall call a *Lambert Quadrilateral*, has an important part to play in later developments.

Theorem 1. In a trirectangular quadrilateral the fourth angle is acute.

Let $ABCD$ (Fig. 33) be a Lambert Quadrilateral with the angles at A, B and D right angles. We wish to prove that the angle at C is acute.

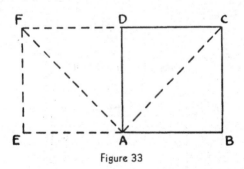

Figure 33

Produce BA through A to E so that AE is equal to BA. At E draw EF perpendicular to BE and equal to BC. Join F to A and D, and draw AC. From the congruence of right triangles FEA and CBA follows the congruence of triangles FAD and CAD. Then angle FDA is a right angle, the points F, D and C are collinear, and $EBCF$ is a Saccheri Quadrilateral. Thus the angle at C is acute.

A useful theorem in regard to a more general quadrilateral, with only two right angles, may very well be inserted here. From it come immediately some important properties of the Saccheri and Lambert Quadrilaterals.

Theorem 2. If, in the quadrilateral $ABCD$ (Fig. 34), the angles at two consecutive vertices A and B are right angles, then the angle at C is larger than or smaller than the angle at D according as AD is larger than or smaller than BC, and conversely.

If AD is larger than BC, measure off on AD the segment AE equal to BC and draw EC. Then $ABCE$ is a Saccheri Quadrilateral with angles AEC and BCE equal. Since
$$\angle BCD > \angle BCE$$
and
$$\angle AEC > \angle ADC,$$
then
$$\angle BCD > \angle ADC.$$
In the same way, if AD is smaller than BC, angle BCD can be shown to be smaller than angle ADC.

The proof of the converse by *reductio ad absurdum* is left as an exercise for the reader.

EXERCISES

1. If the angles at A and B (Fig. 34) are right angles and the angles at C and D are equal, prove that the figure is a Saccheri Quadrilateral.

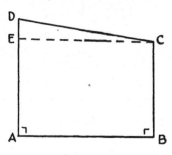

Figure 34

2. Prove that, in a Lambert Quadrilateral, the sides adjacent to the acute angle are greater than their respective opposite sides.

3. Which is the greater, the base or the summit of a Saccheri Quadrilateral?

4. Prove that, if perpendiculars are drawn from the extremities of one side of a triangle to the line passing through the midpoints of the other two sides, a Saccheri Quadrilateral is formed. As a consequence, prove that the perpendicular bisector of any side of a triangle is perpendicular to the line joining the midpoints of the other two sides.

5. Prove that the segment joining the midpoints of two sides of a triangle is less than one-half the third side.[4]

6. Show that a line through the midpoint of one side of a triangle perpendicular to the line which bisects a second side at right angles bisects the third side.

7. Prove that the line joining the midpoints of the equal sides of a Saccheri Quadrilateral is perpendicular to the line joining the midpoints of the base and summit and that it bisects the diagonals.

44. The Sum of the Angles of a Triangle.

Theorem 1. The sum of the angles of every right triangle is less than two right angles.

Let ABC (Fig. 35) be any right triangle with the right angle at C.

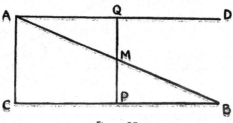

Figure 35

We know that each of the other angles is acute since the sum of two angles of a triangle is always less than two right angles. At A construct angle BAD equal to angle ABC. From the midpoint M of AB draw MP perpendicular to CB. P will lie between B and C. On AD cut off AQ equal to PB and draw MQ. Then triangles MBP

[4] W. H. Young has assumed this and used it as the characteristic postulate in a development of Hyperbolic Geometry. On the assumptions that the segment is equal to or greater than the third side follow Parabolic and Elliptic Geometries, respectively. See the Quarterly Journal of Pure and Applied Mathematics, Vol. XLI, 1910, pp. 353-363, and the American Journal of Mathematics, Vol. 33, 1911, pp. 249-286.

and *MAQ* are congruent and it follows that angle *AQM* is a right angle, that points *Q*, *M* and *P* are collinear, and consequently that *ACPQ* is a Lambert Quadrilateral with acute angle at *A*. Thus the sum of the acute angles of the right triangle *ABC* is less than one right angle and the sum of all three angles is less than two right angles.

Theorem 2. The sum of the angles of every triangle is less than two right angles.

The theorem has already been proved for right triangles, so we assume that triangle *ABC* (Fig. 36) has none of its angles a right angle. Since at least two of the angles of every triangle are acute, we may assume that the angles at *B* and *C* are acute and draw the altitude *AD* from *A* to *BC*, *D* falling between *B* and *C*.

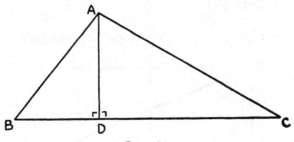

Figure 36

Thus triangle *ABC* is divided into two right triangles *ADB* and *ADC*. Since the sum of angles *ABD* and *BAD* is less than one right angle, as is also the sum of angles *ACD* and *CAD*, the sum of the angles of triangle *ABC* is less than two right angles.

The difference between two right angles and the angle-sum of a triangle is called the *defect* of the triangle.

Corollary. The sum of the angles of every quadrilateral is less than four right angles.

Theorem 3. If the three angles of one triangle are equal, respectively, to the three angles of a second, then the two triangles are congruent.

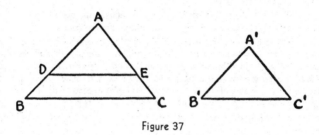

Figure 37

Let angles *A*, *B*, *C* of triangle *ABC* (Fig. 37) be equal, respectively, to angles *A'*, *B'*, *C'* of triangle *A'B'C'*. If any pair of corresponding sides, say *AB* and *A'B'*, are equal, the triangles are of course congruent. Assume that *AB* and *A'B'* are unequal; then one of them, say *AB*, is the larger. Cut off on *AB* the segment *AD* equal to *A'B'* and on *AC* segment *AE* equal to *A'C'*. That *AE* is shorter than *AC* will be verified presently.

Since triangles *ADE* and *A'B'C'* are congruent, it is evident that *BCED* is a quadrilateral with the sum of its angles equal to four right angles. But this is impossible and consequently *AB* and *A'B'* must be equal and the triangles congruent.

If *AE* were equal to *AC*, angles *BCA* and *DCA* would have to be equal. But this is impossible if *AD* is less than *AB*. If *AE* were greater than *AC*, a situation would be encountered in which an exterior angle of a triangle would be equal to one of the opposite interior angles, but this again is absurd.

Thus we reach the remarkable conclusion that in Hyperbolic Geometry similar triangles, or indeed similar polygons, of different sizes do not exist.

We shall show in Section 54 how to construct a triangle, given the three angles.

EXERCISES

1. Using Theorem 2, modify the proof of Lemma 2, Section 13, to show that, if a line through a given point and cutting a given line revolves about the given point and approaches parallelism to the given line, then the angle which it forms with the given line approaches zero. In other words, show how two parallel lines may be thought of as intersecting at a zero angle.

2. Prove that two Saccheri Quadrilaterals with equal summits and equal summit angles, or with equal bases and equal summit angles, are congruent.

3. A segment joining a vertex of a triangle to a point on the opposite side is called a transversal.[5] A transversal divides a triangle into two subtriangles, and one or both of these can be subdivided by transversals, and so on. Prove that, if a triangle is subdivided by arbitrary transversals into a finite number of triangles, the defect of the triangle is equal to the sum of the defects of the triangles in the partition. What does this suggest about the defect of a triangle as compared to its size?

4. Prove that the angle-sum of a polygon of n sides is less than $(n - 2)$ times two right angles.

45. The Common Perpendicular of Two Non-Intersecting Lines.

We next turn our attention to non-intersecting lines. If two lines are perpendicular to the same line, they are non-intersecting. The converse statement is also true and describes one of the most striking properties of non-intersecting lines.

Theorem. Two non-intersecting lines have one and only one common perpendicular.

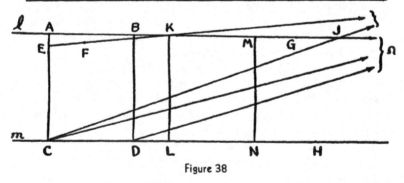

Figure 38

Let l and m (Fig. 38) be any pair of non-intersecting lines. Select any two points A and B on l and draw the perpendiculars AC and BD

[5] See Hilbert, *The Foundations of Geometry*, translated by E. J. Townsend, p. 63 (Chicago, 1902) or *Grundlagen der Geometrie*, 5th edition, p. 58 (Leipzig and Berlin, 1922).

to *m*. If *AC* and *BD* are equal, *ABDC* is a Saccheri Quadrilateral, from which it follows at once that *l* and *m* have a common perpendicular. If *AC* and *BD* are not equal, let us assume that *AC* is the longer and on it measure off *CE* equal to *BD*. At *E* draw *EF* on the side of *AC* on which *B* and *D* lie, making angle *CEF* equal to angle *DBG*, where *G* is any point on *l* on the side of *B* opposite *A*.

We propose to show that *EF* will cut *l*, if sufficiently produced. To do this, draw[6] *CΩ* and *DΩ* parallel to *l* in the sense *AB*. These lines must lie within the angles *ACH* and *BDH*, respectively, where *H* is any point on *m* on the side of *D* opposite *C*. Since angle *HDΩ* is greater than *HCΩ*, a line *CJ* drawn through *C*, making with *CH* the same angle which *DΩ* makes with it, will cut *l* in a point *J*. A comparison of figures *FECJ* and *GBDΩ* convinces us that *EF* is parallel to *CJ* and consequently must intersect side *AJ* of triangle *ACJ* in a point *K*.

Draw *KL* perpendicular to *m*. On *l* and *m*, respectively, on the side of *BD* opposite *AC*, cut off *BM* equal to *EK* and *DN* equal to *CL* and draw *MN*. By means of congruent triangles, it is easy to show that quadrilaterals *EKLC* and *BMND* are congruent. As consequences, *MN* is perpendicular to *m*, and *MN* and *KL* are equal. The line joining the midpoints of summit and base of the Saccheri Quadrilateral *KMNL* is a common perpendicular to *l* and *m*.

There cannot be more than one such common perpendicular, for if there were two, there would exist a quadrilateral with an angle-sum of four right angles. But this is absurd.

We call attention to the fact that the argument[7] above not only proves the existence of a unique common perpendicular to two non-intersecting lines, but, assuming that we can construct the parallels to a line through a given point, supplies a method of constructing this perpendicular when two such lines are given.

46. Ultra-Ideal Points.

In Hyperbolic Plane Geometry, two lines intersect, are parallel or are non-intersecting. Already we have adopted[8] a convenient

---

[6] We are merely employing here the phraseology customarily used in describing a drawing. We have not yet shown how to construct a parallel to a line through a point, but the proof does not depend upon our ability to do so.

[7] Due to Hilbert, *loc. cit.*, p. 164.

[8] See Section 39.

terminology with reference to parallels, one which allows us on occasion to regard them as intersecting. The time has come to extend these concepts to include non-intersecting lines.

As the name implies, two non-intersecting lines do not have a point in common. But they do have *something* in common; they have a common perpendicular. Arguing in much the same fashion as that followed in introducing ideal points, we choose to recognize this relationship by saying that two non-intersecting lines have in common, or intersect in, an *ultra-ideal point*. Thus all of the lines perpendicular to a given line will be regarded as being concurrent in an ultra-ideal point and constituting a sheaf of lines with an ultra-ideal vertex. Two given non-intersecting lines then determine an ultra-ideal point, and the sheaf of lines having this point for vertex consists of all of the lines cutting at right angles the common perpendicular to the two given lines. Corresponding to every ultra-ideal point is a line, its *representative line*, such that every line perpendicular to it passes through the point; corresponding to every line is an ultra-ideal point through which pass all of the perpendiculars to the line. We shall follow the convention of designating ultra-ideal points by large Greek letters (generally Γ) with subscripts designating the representative lines. Thus Γ_l denotes the ultra-ideal point through which pass all of the lines perpendicular to the line l.

The nature and significance of this new viewpoint, under which non-intersecting lines are regarded as intersecting, should be reasonably clear to the reader by this time. Many of the general remarks made in the introduction of ideal points hold also for ultra-ideal points.

EXERCISE

Prove that every line contains an infinite number of ultra-ideal points.

47. The Variation in the Distance between Two Lines

We wish now to determine the effect which a change in the position of a point on a line will have upon the length of the perpendicular from the point to another line. There are three cases to be considered inasmuch as two lines may be intersecting, parallel or non-intersecting.

Theorem 1. Two intersecting lines diverge continuously from their point of intersection, the perpendicular distance from a point on one of them to the other increasing without limit as the point moves away from the point of intersection, and becoming smaller than any assigned distance, however small, as it moves toward it.

Let l and m (Fig. 39) be any two non-perpendicular lines intersecting at O, and let P_1 and P_2 be any two points on l, on the same side of O, and so situated that OP_2 is greater that OP_1. Draw P_1Q_1 and P_2Q_2 perpendicular to m. Then, in the birectangular quadrilateral $P_1P_2Q_2Q_1$, angle $P_2P_1Q_1$ is obtuse and angle $P_1P_2Q_2$ is acute; therefore P_2Q_2 is greater than P_1Q_1. Thus we have proved that the perpendicular from a point on one of two intersecting lines to the other increases as the point moves away from the intersection and

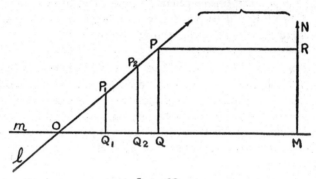

Figure 39

decreases as it moves toward it. To complete the proof we need to show that the point can be so chosen that the distance will be equal to, and hence greater than or less than, any assigned length. This amounts to proving that, given one acute angle of a right triangle and the length of the opposite side, no matter how large or small, the triangle can be constructed. We shall prove this shortly and may assume it here. The reader should observe that this construction cannot be made in Hyperbolic Geometry as simply as in Euclidean.

It will, however, be instructive to show without delay, and in a different way, that the distance becomes greater than any assigned segment as the point moves away from O. In order to do this, we recall that to every acute angle, regarded as an angle of parallelism, there is a corresponding distance. A little later we shall devise a construction for the distance when the angle is given. Regarding angle P_1OQ_1 (Fig. 39) as an angle of parallelism, measure OM on m equal to the corresponding distance and draw MN perpendicular to m at M. Then MN is parallel to l. To show that the perpendicular distance to m from a point on l becomes longer than any assigned segment, however long, cut off MR on MN equal to the given segment, and draw the perpendicular to MN at R. It is easy to see that this perpendicular will cut l in a point P, and that the perpendicular PQ from P to m is greater than RM and hence longer than the given segment. One should notice that, while a perpendicular can be drawn to m from every point of l, if a perpendicular is drawn to m at any point Q, this perpendicular will cut l only so long as OQ is less than OM, will be parallel to it when Q coincides with M, and cease to cut it at all when OQ is greater than OM.

Theorem 2. Two parallel lines converge continuously in the direction of parallelism and diverge continuously in the opposite direction, the perpendicular distance from a point on one of them to the other becoming smaller than any assigned distance, however small, as the point moves in the direction of parallelism, and larger than any assigned distance, however large, as it moves in the opposite direction.

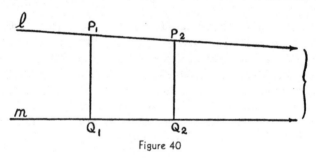

Figure 40

Let l and m (Fig. 40) be any two parallel lines and let P_1 and P_2 be any two points on l, P_2 lying on the side of P_1 in the direction of parallelism. Since, in the birectangular quadrilateral $P_1P_2Q_2Q_1$, the angle at P_1 is smaller than the angle at P_2, P_2Q_2 is shorter than P_1Q_1. All that is needed now is to show that a point can always be found on l having any given perpendicular distance to m, however large or small. To do this, draw from any point P (Fig. 41) on l the perpendicular PQ to m. If PQ is equal to the given distance, the required point has been found. Otherwise lay off on QP, or QP produced, the segment QR equal to the given distance. Through R draw the parallel to m in the sense opposite that in which l is drawn. It is readily shown that this parallel, produced in the direction opposite that of parallelism if necessary, is cut by l in a point S.

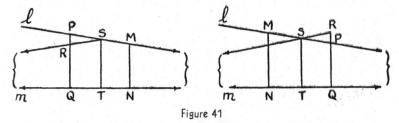

Figure 41

Draw ST perpendicular to m. On l and m measure off SM and TN equal to SR and TQ, respectively, on the side of ST opposite PQ. Join M and N. There will now be no difficulty in proving that MN is perpendicular to m and equal to the given distance.

Thus it is clear that parallel lines are not equidistant as in Euclidean geometry. Since the distance from a point on one of two parallel lines to the other approaches zero as the point moves in the direction of parallelism, and the lines do not intersect in the strict sense, they are *asymptotic*. The reader is also in a position now to recognize one of the most striking of the peculiar features of Hyperbolic Geometry: *two pairs of parallel lines are always congruent*.

Theorem 3. Two non-intersecting lines continuously diverge on each side of their common perpendicular, the perpendicular distance from a point on one to the other being shortest when measured along that perpendicular and increasing as the point moves away from the perpendicular in either direction, becoming larger than any assigned distance, however large.

Let l and m (Fig. 42) be any two non-intersecting lines and MN their common perpendicular. Let P_1 and P_2 be any two points on l

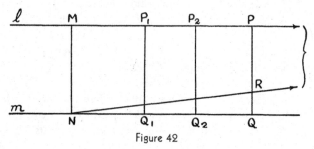

Figure 42

on the same side of M, MP_2 being greater that MP_1. Draw P_1Q_1 and P_2Q_2 perpendicular to m. Examination of the Lambert Quadrilaterals MP_1Q_1N and MP_2Q_2N shows that both P_1Q_1 and P_2Q_2 are larger than MN and that angles MP_1Q_1 and MP_2Q_2 are acute. Then in the birectangular quadrilateral $P_1P_2Q_2Q_1$ the angle at P_1 is larger than that at P_2 and consequently P_2Q_2 is larger than P_1Q_1. Thus as a point moves along l, away from M, the perpendicular distance to m increases.

That this distance increases continuously and becomes larger than any assigned distance, however large, will follow presently when we prove that a unique Lambert Quadrilateral, such as MP_1Q_1N, can always be constructed when MN and P_1Q_1 are given, regardless of how long P_1Q_1 is, provided, of course, that it is longer than MN. But we need not wait for this to show that the distance becomes larger without limit. Let P (Fig. 42) be any point on l and draw PQ perpendicular to m. Through N draw the parallel to l on the side of MN on which P lies. By joining P to N, it can be seen that this parallel cuts PQ in a point R, so that PQ is greater than RQ. But RQ is the perpendicular distance from a point on one of two intersecting lines to the other. As P moves away from M, R moves away from N, RQ becomes large without limit and hence so also does PQ.

48. The Perpendicular Bisectors of the Sides of a Triangle.

In Hyperbolic Geometry, as is the case in Euclidean, the perpendicular bisectors of the sides of a triangle are concurrent and so also are the bisectors of the angles, the altitudes and the medians. Here, however, the lines must at times be regarded as intersecting in ideal or ultra-ideal points. The proofs for these concurrence theorems are, on the whole, not so easily obtained as in Euclidean Geometry. Some of the difficulty encountered is rather fundamental in character. We shall treat here only one of these theorems, one which we shall utilize very soon.

Theorem. The perpendicular bisectors of the sides of a triangle are concurrent.

There are three cases to be considered.

CASE I. If the perpendicular bisectors of two of the sides of a triangle intersect in an ordinary point, then it can be proved by congruence theorems, just as in Euclidean Geometry, that the perpendicular bisector of the third side passes through this point.

CASE II. If the perpendicular bisectors of two of the sides of a triangle are non-intersecting, then they will have a common perpendicular and intersect in an ultra-ideal point. We shall prove that the perpendicular bisector of the third side is perpendicular to this common perpendicular also, that is, that it passes through the same ultra-ideal point.

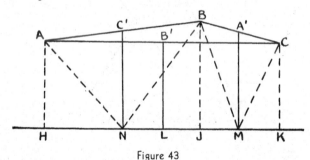

Figure 43

Let ABC (Fig. 43) be the triangle, with A', B', C' the midpoints of the sides opposite A, B, C, respectively. Let the perpendiculars to sides AB and BC at C' and A' be non-intersecting. They then have a common perpendicular MN. We are to prove that the perpendicular to AC at B' is also perpendicular to MN.

Construct AH, BJ, CK and $B'L$ perpendicular to MN. Then, if lines AN, BN, BM and CM are drawn, it is easy to prove that AH and CK are each equal to BJ and hence equal to one another. Thus $AHKC$ is a Saccheri Quadrilateral and the line $B'L$ through B', the midpoint of the summit, perpendicular to the base, is perpendicular to the summit also. This proves that the perpendicular bisectors of the three sides of the triangle have in this case an ultra-ideal point in common.

CASE III. Finally, if the perpendicular bisectors of two of the sides of a triangle are parallel, then the perpendicular bisector of the third side must be parallel to each of them. For if it intersects either, or is non-intersecting with respect to either, a contradiction is encountered owing to what has been proved above. The only thing which is left to be determined is whether or not the two parallel bisectors are parallel to the third in the same sense or in opposite senses. We shall show that the latter is impossible.

If the two perpendicular bisectors which are parallel to one another are parallel to the third in opposite senses, they will form, we may say, a triangle $\Omega_1\Omega_2\Omega_3$ (Fig. 44) with its vertices ideal points.

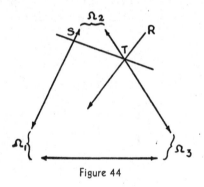

Figure 44

Now no straight line can cut all three sides of such a triangle in ordinary points. If ST, for example, cuts $\Omega_1\Omega_2$ in S and $\Omega_2\Omega_3$ in T, it will be made apparent by drawing $T\Omega_1$ and producing it in the opposite sense to any point R, that ST will lie within the vertical angles $\Omega_2T\Omega_1$ and $RT\Omega_3$ and consequently be non-intersecting with regard to

$\Omega_1\Omega_3$. But there is always one line, at least, which intersects all three perpendicular bisectors of the sides of every triangle, as we shall show.

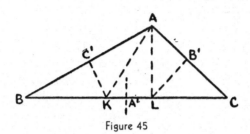

Figure 45

Assume that the triangle ABC (Fig. 45) has no two of its angles equal and let the angle at A be the greatest. Construct angle BAK equal to the angle at B and angle CAL equal to the angle at C. Then it follows readily that the perpendicular bisectors of all three sides of the triangle intersect side BC. The proof can be modified for the cases where two or all three of the angles are equal.

Thus we conclude that, if the perpendicular bisectors of two sides of a triangle intersect in an ideal point, the perpendicular bisector of the third side must pass through this ideal point.

EXERCISES

1. Prove that the internal bisectors of the angles of a triangle are concurrent.

2. Show that the perpendicular bisectors of the sides of a triangle are the altitudes of the triangle having for vertices the midpoints of the sides of that triangle. Then prove that the altitudes of a triangle are concurrent, provided the perpendiculars to two of the altitudes, at the vertices from which they are drawn, intersect in an ordinary point.

49. The Construction of the Parallels to a Line through a Point.

As a consequence of the characteristic postulate of Hyperbolic Geometry, we know that through a given point P two parallels can be drawn to a given line l. Thus far we have made no attempt to show how these parallels can actually be constructed. In order to do this we shall need a lemma.

Lemma.[9] The midpoints of the segments joining pairs of corresponding points of two congruent point rows lie on a straight line, unless these segments have a common midpoint.

Let $ABC \ldots$ and $A'B'C' \ldots$ (Fig. 46) be two congruent point rows such that $AB = A'B'$, $BC = B'C'$, etc., and let M, N, P be the midpoints of AA', BB', CC', respectively. It is easy to see that if M coincides with N then P does also, and all of the segments

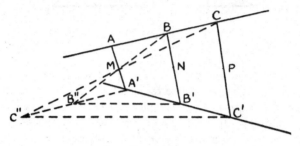

Figure 46

have the same midpoint. Consequently assume M, N and P distinct. Draw BM and produce to B'' so that MB'' is equal to BM. Then B'' and B' are distinct points. When B'' is joined to B' and also to A', it is clear that the perpendicular bisector of the base $B''B'$ of the isosceles triangle $B''A'B'$ is also the perpendicular bisector of the base of the triangle $B''BB'$, and hence is perpendicular to MN (Exercise 4, Section 43). Next, produce $A'B''$ to C'' so that $B''C''$ is equal to BC and draw $C''C'$, $C''M$ and MC. By the use of congruent triangles one can easily show that C'', M and C are collinear and that M is the midpoint of CC''. Then the perpendicular bisector of the base $C''C'$ of the isosceles triangle $C''A'C'$ is also the perpendicular bisector of the base of triangle $C''CC'$ and hence is perpendicular to MP. But the perpendicular bisectors of the bases of the isosceles triangles $B''A'B'$ and $C''A'C'$ are the same line. Consequently MN and MP are perpendicular to the same line and M, N and P must be collinear.

[9] J. Hjelmslev, *Neue Begründung der ebene Geometrie, Mathematische Annalen*, Vol. 64, 1907, p. 449.

Corollary. If the lines supporting the congruent point-rows are parallel, the locus of the midpoints is parallel to them in the same sense.

With the lemma proved, we are ready to return to the fundamental construction problem.

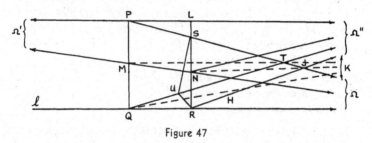

Figure 47

If l (Fig. 47) is any given line and P is any given point not on that line, we wish to construct the two parallels through P to l. Let us confine our attention to one of them, say the right-hand parallel $P\Omega$, and, since we know it exists, assume that it has been drawn. First construct the perpendicular PQ from P to l. Then select any point R on l on the side of Q in the direction of parallelism, cut off PS on $P\Omega$ equal to QR and draw SR. Connect M and N, the midpoints of PQ and RS, respectively. As a consequence of the lemma and its corollary, we know that MN is parallel to l and PS. Draw the line $\Omega'P\Omega''$ through P, perpendicular to PQ. If NM is produced through M it is obvious that it is parallel to $P\Omega'$. Construct through Q the right-hand parallel to $P\Omega''$; it must intersect $P\Omega$ in a point T. Measure off QU on $Q\Omega''$ equal to QR and PS, and draw SU and RU. Since triangles PTQ and STU are both isosceles, MT is perpendicular to SU at its midpoint and is also perpendicular to line $\Omega K\Omega''$ to which $T\Omega$ and $T\Omega''$ are the two parallels through T. Furthermore, since triangle UQR is isosceles, the perpendicular bisector of UR bisects angle $\Omega''Q\Omega$ and consequently also is perpendicular to $\Omega K\Omega''$. It follows from the results of the preceding section that the line perpendicular to side SR of triangle SUR at its midpoint is perpendicular to $\Omega K\Omega''$. This perpendicular to SR at N

then bisects the angle formed by $N\Omega$ and the line $N\Omega''$ drawn through N parallel to $P\Omega''$. Hence SR bisects angle $\Omega'N\Omega''$, intersects $\Omega'P\Omega''$ at a point L and is perpendicular to it at that point. These results may be summarized as follows:

Construction.[10] To construct the lines parallel to a given line l through a given point P not on the line, draw the perpendicular PQ from P to l and measure off on l in either direction any distance QR. At P draw the line PL perpendicular to PQ and construct the perpendicular RL from R to PL. With P as a center and radius equal to QR, draw an arc of a circle cutting LR at S. PS is one of the parallels to l through P. The other is obtained by constructing R on the opposite side of Q.

That the arc thus described will actually intersect segment LR in one and only one point follows from the fact that a unique parallel exists in each direction and that we have proved PS equal to QR.

Before turning to other matters, we wish to point out another property of Figure 47 for which we shall have use presently. If RH is drawn through R making angle SRH equal to angle $RS\Omega$, it is easy to show that RH and $S\Omega$ will meet in a point J through which will pass the perpendicular bisector of SR. But, since the latter is perpendicular to $\Omega K\Omega''$ and bisects angle SJR, RJ must be the other parallel to $\Omega K\Omega''$ through J and thus be parallel to $P\Omega''$. Therefore LR is the distance corresponding to angle LSP regarded as an angle of parallelism.

EXERCISE

Use the lemma to prove that the line joining the midpoints of the base and summit of a Saccheri Quadrilateral, the line joining the midpoints of the other two sides and the line joining the midpoints of the diagonals are concurrent. (Compare with Exercise 7, Section 43.)

---

[10] This construction is the one given by Bolyai in paragraph 34 of the famous *Appendix*. The proof, however, is due to Liebmann, *Nichteuklidische Geometrie*, pp. 35–37, 2nd edition (Berlin and Leipzig, 1912).

50. The Construction of a Common Parallel to Two Intersecting Lines.

It is probably not necessary to remark that, by the construction described in the last section, one is able to construct the angle of parallelism corresponding to any given distance; in other terms, if a segment d is given, $\Pi(d)$ can be constructed. This directs our attention to the reverse construction: given $\Pi(d)$ to construct d, or, what is the same thing, to construct a line perpendicular to one of two intersecting lines and parallel to the other. We shall find it convenient to discover first how to draw a common parallel to two intersecting lines.

Let l and m (Fig. 48) be any two intersecting lines, O their point

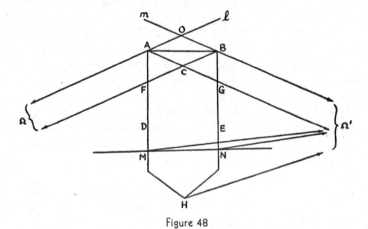

Figure 48

of intersection and $\Omega O \Omega'$ any one of the four angles which they form. Select points A and B on $O\Omega$ and $O\Omega'$, respectively, so that OA and OB are equal, then draw AB and the parallels $A\Omega'$ and $B\Omega$. The latter two lines will intersect in a point C. Construct next the bisectors AD and BE of angles $\Omega A\Omega'$ and $\Omega B\Omega'$, which will cut $B\Omega$ and $A\Omega'$, respectively, in points F and G. From a comparison of figures $AO\Omega'$ and $BO\Omega$, it follows that angles OAC and OBC are equal, and consequently so also are angles ΩAC and $CB\Omega'$. We are now prepared to show that AD and BE are non-intersecting.

First let us assume that AD and BE intersect in a point H. It is easy then to show that angles BAH and ABH are equal and therefore also the segments AH and BH. Combining this result with the fact that angles $HA\Omega'$ and $HB\Omega'$ are equal, we conclude that, if $H\Omega'$ is drawn, angles $AH\Omega'$ and $BH\Omega'$ will be equal. But this is impossible, and accordingly AD and BE do not intersect. That they do not intersect if produced through A and B follows from the fact that the sum of angles BFD and FBE is less than two right angles.

Next we assume that AD and BE are parallel. In this case, by comparing the figure formed by the two parallels $A\Omega$ and $F\Omega$ and the transversal AF with that formed by the two parallels FD and BE and the transversal FB, we easily prove that AF and FB are equal, since angles ΩAF and FBE are equal as well as angles $AF\Omega$ and BFD. Then angle BAF is equal to angle ABF. But this is absurd, and therefore AD and BE are not parallel, at least in the direction considered. But if they are parallel at all, it must be in that direction.

Since AD and BE are neither intersecting nor parallel, they must have a unique common perpendicular. Let it be the line MN. We shall prove that MN is parallel to $O\Omega$ and $O\Omega'$.

Inspection of the quadrilateral $AMNB$ reveals that segments AM and BN are equal. If MN is not parallel to $O\Omega'$, draw $M\Omega'$ and $N\Omega'$. Comparing figures $MA\Omega'$ and $NB\Omega'$, we see that under these circumstances angles $AM\Omega'$ and $BN\Omega'$ are equal. But as a consequence we are led to the absurd conclusion that figure $MN\Omega'$ has an exterior angle equal to the opposite interior angle. Then MN is parallel to both l and m, and it is clearly unique for the directions indicated on the lines, regardless of the length of OA and OB. This method not only proves the existence of a common parallel but supplies a mode of effecting its construction. As a matter of fact, two intersecting lines always have four common parallels.

But two lines do not need to be intersecting in order to have a common parallel. A common parallel can be constructed for any two lines. If they do not intersect, one has only to draw through any point of the first a parallel to the second and then construct a common parallel to the pair of intersecting lines. Obviously two parallel lines have only one common parallel which is parallel to them in opposite directions.

Finally, interpreted otherwise, this construction of a line parallel to two lines in given senses enables us to effect the construction of the line joining any two given ideal points.

51. The Construction of a Line Perpendicular to One of Two Intersecting Lines and Parallel to the Other.

We return now to the problem of constructing a line perpendicular to one of two intersecting lines and parallel to the other, or, otherwise, of constructing the distance corresponding to any acute angle regarded as an angle of parallelism. This construction is readily accomplished by the use of the results of the last section.

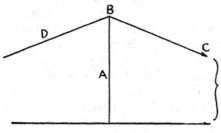

Figure 49

Given the acute angle ABC (Fig. 49), we wish to construct a line perpendicular to BA and parallel to BC. All that is needed is to construct the angle ABD equal to angle ABC. Then the common parallel to BC and BD will be perpendicular to BA and parallel to BC. This construction can always be made, whatever the size of the given acute angle, no matter how small or how near a right angle.

Here again attention is called to the generality of the construction. A line can be constructed perpendicular to one of two lines and parallel to the other even when they do not intersect, whether they be parallel or non-intersecting. The modification of the construction for these cases has already been suggested.

EXERCISE

How many lines can be constructed which are perpendicular to one of two given lines and parallel to the other, if the given lines (a) intersect at an acute angle? (b) are perpendicular? (c) are parallel? (d) are non-intersecting?

52. Units of Length and Angle.

We speak of the units which we use for measuring angles as being *absolute* because these units are conveniently chosen parts of a straight angle or right angle. The latter angles are fundamental figures; they can be constructed at will and do not vary in size. There is no need of preserving in any bureau of standards a right angle. Anybody can construct one, whenever he wishes, with great precision.

The measurement of angles does not depend upon the theory of parallels. What we have said holds for angles in all three geometries. The student has been accustomed to use as the unit of angle the one obtained by dividing a straight angle by π. We shall find it convenient to adopt this unit in what follows. It should be clearly understood that π is here merely the familiar abstract number which is perhaps most easily described as four times the limit of the sum of the series $1 - 1/3 + 1/5 - 1/7 + 1/9 \ldots \ldots$, and has for an approximate value 3.1416. It is to be regarded in no other light. Emphatically it is not the ratio of the length of the circumference of a circle to its diameter. That constant ratio occurs in Euclidean Geometry as a consequence of the parallel postulate. In Hyperbolic Geometry this ratio is, as we shall discover, variable. As a matter of fact, the unit angle just designated does not have here the simple geometric interpretation of Euclidean Geometry.

On the other hand, we speak of the units of length used in Euclidean Geometry as *relative*. They are arbitrary and conventional. There is no fundamental length which can be constructed, expunged, and then reconstructed, to be used in whole or in part as a unit, without the necessity of reference to some preserved standard. The units, if they are not to vary as the decades go by, must be guarded by a bureau of standards or some other agency.

Now, however, we are prepared to understand why, on the contrary, units of length are absolute in Hyperbolic Geometry. Corresponding to every segment of line, however large or small, there is a unique acute angle, its angle of parallelism, and conversely. The one-to-one correspondence between line segments and acute angles enables us to designate any segment by referring to its associated angle. Thus we could, if we wished, choose as our unit of length the distance corresponding to the angle $\pi/4$ as angle of parallelism.

The angle can easily be constructed and thus, theoretically at least, the unit of line.

The angle cannot, it goes without saying, be used directly as a measure of the segment, for it does not vary in proportion to it. Indeed, as the segment increases the angle decreases. In the next chapter we shall actually succeed in expressing the distance as a function of the angle. Our unit of distance will then correspond to the angle which makes the function equal to unity.

53. Associated Right Triangles.

We are now in a position to obtain a very important result in the theory of right triangles. But first, in order to avoid confusion, let us adopt a standard notation.

Let ABC (Fig. 50) be any right triangle with C as the vertex of the right angle. Designate by λ and μ the measures of the angles at A and B and by a, b, c, the measures of the sides opposite the vertices A, B, C, respectively. We shall represent angles $\Pi(a)$, $\Pi(b)$ and $\Pi(c)$, respectively, by α, β, and γ. The distances corresponding to the angles λ and μ, regarded as angles of parallelism, will be denoted by l and m, so that λ is $\Pi(l)$ and μ is $\Pi(m)$. The complements of angles α, β, γ, λ, μ are conveniently symbolized by α', β', γ', λ', μ', and, as they are acute angles, they have corresponding distances which we designate by a', b', c', l', m', so that, for example, the sum of $\Pi(a)$ and $\Pi(a')$ is a right angle.

A right triangle can be uniquely constructed in Hyperbolic Geometry when angle μ and hypotenuse c are given. The construction,

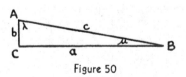

Figure 50

which involves merely the drawing of the perpendicular from a point to a line, is simple and, as a matter of fact, is the one used in Euclidean Geometry. There are no restrictions as to sizes or relative sizes of μ and c, except that the angle must be acute.

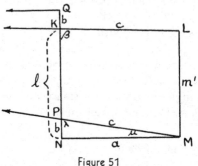

Figure 51

Starting with any right triangle *ABC* (Fig. 50), let us construct a Lambert Quadrilateral *KLMN* (Fig. 51), acute angle at *K*, with *KL* equal to *c* and *LM* equal to *m'*. This can always be done uniquely by drawing first a right angle *KLM*, measuring *LK* equal to *c*, *LM* equal to *m'*, constructing the perpendicular *MN* to *LM* at *M* and then drawing the perpendicular *KN* from *K* to *MN*. With *M* as center and radius equal to *c* strike an arc of a circle cutting *KN* at *P*. We know that this arc will cut segment *KN* owing to what was proved in Section 49. Indeed, *MP* is parallel to *LK* and angle *PML* is equal to μ', that is, it is the angle of parallelism for the distance *m'*. Then angle *PMN* is equal to μ since angle *LMN* is a right angle. In other words, right triangles *PMN* and *ABC* are congruent, *NM* and *PN* are equal to *a* and *b*, respectively, and angle *NPM* is equal to λ. Furthermore, from the remarks made at the close of Section 49, we infer that *KN* is equal to *l*. Finally, if *NK* is produced through *K* to *Q*, so that *KQ* is equal to *b*, the perpendicular drawn to *KQ* at *Q* will be parallel to *MP*, since *PQ* is equal to *l*. It follows that angle *NKL* is β, the angle of parallelism for the distance *b*.

We summarize the results obtained so far as follows: Given a right triangle with parts *a*, *b*, *c*, λ, μ (Fig. 50), there can always be constructed uniquely a Lambert Quadrilateral *KLMN* (Fig. 51) with parts *a*, β, *c*, *l*, *m'*, and conversely. In other words, the existence of, or possibility of constructing, one implies the existence of, or possibility of constructing, the other.

The reader should observe carefully how to letter the corresponding Lambert Quadrilateral when the triangle is given, and conversely. Once he has learned to do this he will be ready to follow through the sequence of right triangles and Lambert Quadrilaterals now to be described.

As already explained, the existence of the right triangle

$$a, b, c, \lambda, \mu \tag{1}$$

implies the existence of the Lambert Quadrilateral with parts a, β, c, l, m'. One can then construct a quadrilateral like this one but with the sides which include the acute angle interchanged and also those including the right angle which is opposite the acute angle. This Lambert Quadrilateral, with parts m', β, l, c, a, implies the existence of a second right triangle

$$m', b, l, \gamma, \alpha'. \tag{2}$$

If one reverses the order of the parts of this right triangle, it follows that there exists a quadrilateral with parts b, μ', l, a', c'. Reversal of the sides of this quadrilateral leads to a right triangle with parts

$$c', m', a', \lambda, \beta'. \tag{3}$$

In a similar way is proved the existence of a fourth right triangle

$$l', c', b', \alpha', \mu, \tag{4}$$

and a fifth,

$$a, l', m, \beta', \gamma. \tag{5}$$

If the process is continued, the original triangle is obtained next and the cycle is closed. Thus the existence of one right triangle implies the existence of four right triangles associated with it.

It will later prove a great convenience for the reader to be able to write down this series of five associated right triangles as an aid to effecting certain constructions and in modifying certain formulas. This can be done without the necessity of memorizing the parts, or going through the chain of reasoning described above, by the use of the following device.

Place the letters a', b', c, l, m on the sides of a pentagon on the outside as indicated in Figure 52. Then write the same letters in the same cyclic order on the inside, but with each one rotated counterclockwise one place. Starting with right triangle a, b, c, λ, μ, we can pass to a second of the associated right triangles by finding on the outside the letter corresponding to each letter representing a part of the first triangle and replacing it by the one suggested by the corresponding letter on the inside. Thus a is replaced by b, since a'

corresponds to b'; b is replaced by m', since b' corresponds to m; c is replaced by l, since c corresponds to l; λ is replaced by α', since l corresponds to a'; and finally μ is replaced by γ, since m corresponds to c. Thus we obtain the triangle with parts $b, m', l, \alpha', \gamma$. By using

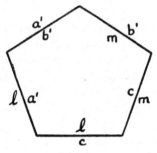

Figure 52

the same drawing we can pass to the third triangle by associating the letters of the *second* with those on the outside and noting the corresponding letters on the inside, and so on until the cycle is completed. Another way is to move each letter on the inside counterclockwise one position and then pass from the *first* triangle to the third, and continuing, from the *first* to the fourth and fifth.

An alternative procedure, which amounts to the same thing, may be preferred. With the letters arranged upon the sides, or *parts*, of the pentagon as on the outside in Figure 52, choose any one of the five parts as *middle part*. Call the sides adjacent to it the *adjacent parts* and the remaining two sides the *opposite parts*. Then to obtain one of the series of associated right triangles, select any one of the five sides as middle part and regard the segment indicated by its letter as hypotenuse. Regard the complementary segments of the segments indicated by the letters on the opposite parts as legs, and the angles of parallelism for the segments indicated by the letters on the adjacent parts as their respective opposite angles. If each part is selected in turn as middle part, all five triangles will be obtained.

As an example of the use that can be made of the right triangles associated with a given right triangle, let us show how to find a point on one of two intersecting lines at a given distance from the other. This problem, suggested previously,[11] is equivalent to that of the construction of a right triangle for which one acute angle and the opposite side are given.

Let a, b, c, λ, μ represent the parts of the right triangle to be constructed with b and μ given. Then we know that there exists a right triangle with parts m', b, l, γ, α'. Since μ is given, μ' and consequently m' can be constructed. Then the construction of the second triangle presents no difficulties, for the lengths of the legs are known. Since one of its acute angles is α', it is a simple matter to draw α and then a. Then with the two legs of the required triangle known, its construction is easily accomplished. There is, by the way, no restriction on the sizes of the given parts in this problem, except that the angle must be acute.

EXERCISES

1. Construct a right triangle, given λ and μ, the two acute angles. Note how the restriction that the sum of λ and μ must be less than a right angle is taken into account.

2. Construct a Lambert Quadrilateral, given the two sides including the acute angle. Can this always be done regardless of the lengths of the given sides?

3. Construct a Saccheri Quadrilateral, given the base and the equal summit angles. Do the same, given the summit and summit angles.

4. Given two non-intersecting lines, find a point on one of them at a given perpendicular distance from the other.[12] Note that this given distance must be greater than the distance between the lines measured along their common perpendicular.

5. Construct a *pseudo-square*, i.e., a quadrilateral with equal angles and equal sides.

54. The Construction of a Triangle when Its Angles are Given.

We now have all that is needed for the construction of a triangle when its three angles are given. This construction[13] can always be made, provided, of course, that the sum of the angles is less than two right angles.

If two of the given angles, or all three of them, are equal, the construction is easy. The problem then resolves itself into the construction of a right triangle given the acute angles.

Assume then that the given angles are unequal. Suppose for the purpose of analysis that the required triangle has been drawn. We

---

[11] See Section 47.

[12] Suggested in Section 47.

[13] Due to Liebmann. See his *Nichteuklidische Geometrie*, 2nd edition, p. 42 (Berlin and Leipzig, 1912).

know that at least two of the angles are acute. It will be convenient to designate them by μ and μ_1 (Fig. 53). If the altitude b is
drawn from the third vertex it will divide the third angle into two
acute angles λ and λ_1 and the opposite side into two segments a and

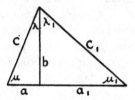

Figure 53

a_1. Let the sides of the third angle be c and c_1. The required triangle is thus divided into two right triangles with parts a, b, c, λ, μ,
and a_1, b, c_1, λ_1, μ_1. From the results of the last section the existence
of these triangles implies the existence of right triangles with parts
c', m', a', λ, β', and c_1', m_1', a_1', λ_1, β'. These triangles PQR and
PQ_1R_1 can be constructed (Fig. 54) and adjoined, the angles λ and λ_1
adjacent to one another and hypotenuse lying on hypotenuse. Then
if QR and Q_1R_1, one or both produced if necessary, intersect in a

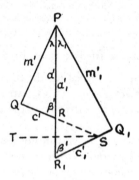

Figure 54

point S, triangle RSR_1 is an isosceles triangle and the bisector ST of
angle RSR_1 is perpendicular to PR_1.

This analysis leads to the following construction. Given the
three angles of a triangle μ, μ_1 and $\lambda + \lambda_1$, the first two acute, con-

struct angle QPQ_1 equal to $\lambda + \lambda_1$ and measure off on its sides PQ equal to m' and PQ_1 equal to m_1'. Draw perpendiculars to PQ and PQ_1 at Q and Q_1, respectively. If these perpendiculars intersect in a point S, draw that bisector of the angles at S such that the perpendicular to it from P lies within the angle QPQ_1. This perpendicular will make with QS an angle of the same size as the one it makes with Q_1S. Call this angle β' and the two angles into which QPQ_1 is divided by the perpendicular λ and λ_1. Once the segment b is constructed, it is mere routine to construct the required triangle.

The construction outlined apparently depends upon the condition that QR and Q_1R_1 intersect. But so far as our investigation has gone, we are not sure that they may not on occasion be parallel or nonintersecting. However, we need not concern ourselves about this, for in any case the perpendicular bisector of RR_1 is the line of symmetry for QR and Q_1R_1. The construction for this line, to be described in Section 57, is perfectly general and may be used to obtain it in any one of the three cases.

55. The Absolute.

Before we turn our attention to other matters, it will be appropriate to add a few remarks regarding ideal and ultra-ideal points. We have already recognized that, when these concepts are taken into account, the statement that two lines determine a point is always true. But it is not always true that two points determine a line and it will be of interest to note the exceptions.

Two ordinary points always determine a line and it is assumed that the line can be drawn. An ordinary point together with an ideal or ultra-ideal point determines a line and the line in each case can be constructed, the construction amounting to drawing a line through a point parallel to a given line in a given sense in one case and perpendicular to a line in the other. We have learned that one and only one line can be drawn parallel to each of two given lines in a specific sense, provided the two lines are not parallel to one another in these senses. Accordingly, two ideal points always determine a line.

But an ideal and an ultra-ideal point do not always determine a line, nor do two ultra-ideal points. In the first case the construction amounts to drawing a line perpendicular to one line and parallel to another. The exception occurs when the given ideal point lies on the line representing the ultra-ideal point. In the second case the construction amounts to drawing a line perpendicular to each of the representative lines of the ultra-ideal points. This can be done only when these lines are non-intersecting. Thus two ultra-ideal points do not determine a line if their representative lines intersect or are parallel.

Our ideas along these lines can be regimented in a remarkable way by means of the accompanying drawing (Fig. 55). The tyro is to

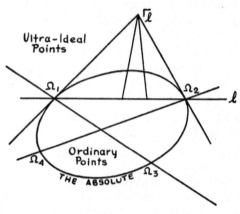

Figure 55

regard it merely as a diagram conveniently exhibiting the relationship between ordinary and extraordinary points.

If any line is allowed to rotate continuously about any one of its points, every point of the line, including each ideal point, will trace out a continuous path. The rotating line will at some time during the complete revolution be parallel in either sense to every line of the plane in each sense. Then each ideal point of the rotating line will eventually coincide with every ideal point in the plane. The path traced out by the ideal points of the rotating line is the locus of all of the ideal points of the plane and this locus we describe as a conic because it has the property that every line of the plane intersects it in two points. It was Cayley[14] who called this locus of the ideal points the *absolute*.

---

[14] See Section 34.

Let the conic in the diagram (Fig. 55) represent the absolute. All points on the conic are then regarded as ideal. All points inside the absolute, that is, all points from which no real tangent lines can be drawn to the conic, represent ordinary points. The points outside are regarded as ultra-ideal. Since every line in Hyperbolic Geometry contains two ideal points, only lines which cut the absolute in two real points are regarded as representing the lines of our geometry. By way of examples, lines $\Omega_1\Omega_3$ and $\Omega_1\Omega_2$ represent parallel lines; lines $\Omega_1\Omega_3$ and $\Omega_2\Omega_4$ are intersecting. The point of intersection Γ_l of the tangent lines to the conic at Ω_1 and Ω_2, where the line l cuts it, is chosen to represent the ultra-ideal point which has l as representative line. All lines through Γ_l are looked upon as perpendicular to l and are non-intersecting. The drawing displays in striking fashion the exceptional cases in which two points do not determine a line.

From our point of view, the absolute is of course inaccessible. We reiterate that the beginner in geometry is to regard the diagram merely as an aid in coördinating ideas and is not to try to make too much out of it. On the other hand, the advanced student, one who understands the viewpoint of projective geometry, will quickly grasp its deep significance and readily derive its implications. He will recognize at once how the character of a geometry depends upon the nature of the absolute.[15]

56. Circles.

We have carried over into the geometry which we are studying, along with other definitions, that of the circle: the locus of points at a constant distance, called the radius, from a fixed point, called the center. But it is probably not necessary to remark that much of the Euclidean theory of the circle must be abandoned or greatly modified here. The careful student will have little difficulty in distinguishing between those Euclidean propositions on the circle which do not remain valid and those which do. For example, the inscribed angle theorems depending on the properties of Euclidean parallels no longer hold; an angle inscribed in a semicircle is no

[15] In regard to the character of the absolute in Euclidean Geometry and — if we are allowed to anticipate — in Elliptic Geometry, see Cayley's *Sixth Memoir*, referred to in Section 34, or consult Veblen and Young: *Projective Geometry*, Vol. II, Ch. VIII (Boston, 1918). See also Coxeter: *Non-Euclidean Geometry* (Toronto, 1942).

longer a right angle or even constant. On the other hand, a line from the center of a circle perpendicular to a chord still bisects the chord; a line perpendicular to a radius at its extremity is still a tangent line.

We do not propose to present here a detailed account of the properties of circles. The reader who is interested may carry out his own investigations. In our limited space we must restrict ourselves to discussion of the broader and more fundamental differences which have been brought about by a change in the postulate of parallels.

A case in point arises when we consider the limiting form of a circle as its radius becomes infinite. In Euclidean Geometry it is a straight line; in Hyperbolic Geometry it is not a line but a curve of peculiar character. We shall find that by skillfully modifying our definition of circle we can, by a simple and natural change of viewpoint, study this curve and its properties without the neccessity of drawing upon the fact that it is the limiting form of a circle. To do this, we introduce in the next section the theory of *corresponding points*.

EXERCISES

1. Prove that if a quadrilateral is inscribed in a circle the sum of one pair of opposite angles is equal to the sum of the other pair.

2. Construct the tangent lines to a given circle from a given point outside the circle.

3. Construct the tangent lines to a given circle parallel to a given line; perpendicular to a given line.

4. Show that an angle inscribed in a semicircle cannot be a right angle, that, as a matter of fact, it must be acute. Prove that if the angle inscribed in a semicircle could be proved constant, it would have to be a right angle and the geometry Euclidean.

57. Corresponding Points.

If two points P and Q, one on each of two straight lines, are so situated that the two lines make equal angles with segment PQ on the same side, then P and Q are called *corresponding points* on the two lines, each one *corresponding* to the other.

In undertaking the task of discussing the elementary properties of corresponding points, we must recognize three cases, since the lines on which the points lie may be parallel, intersecting or non-intersecting.

Corresponding Points on Parallel Lines.

Theorem 1. Given any point on one of two parallel lines, there is always one and only one point on the other which corresponds to it.

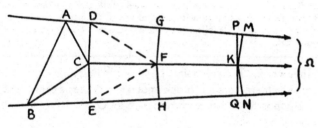

Figure 56

Let $A\Omega$ and $B\Omega$ (Fig. 56) be any two parallel lines with A any point on the first and B any point on the second. Join A and B. If the bisectors of angles $AB\Omega$ and $BA\Omega$ are constructed they will intersect in a point C. It is easy then to prove that the perpendiculars CD and CE, drawn from C to $A\Omega$ and $B\Omega$, respectively, are equal. It follows that $C\Omega$, the common parallel to $A\Omega$ and $B\Omega$ through C, bisects angle DCE, since the angles of parallelism for equal distances are equal. Select any point F on $C\Omega$, or on $C\Omega$ produced through C, and construct the perpendiculars FG and FH to $A\Omega$ and $B\Omega$. Then by means of congruent triangles it can be shown that FG and FH are equal and make equal angles with $F\Omega$. We point out a fact which will be useful later, namely, that DG and EH are also equal.

Having constructed $C\Omega$, draw the perpendicular PK from any point P on $A\Omega$ to $C\Omega$. From K draw the equal perpendiculars KM and KN to $A\Omega$ and $B\Omega$. It is obvious that the point M will lie on the side of P in the direction of parallelism. Then on $B\Omega$ measure off from N, in the direction opposite that of parallelism, the segment NQ equal to MP. Join K and Q. It will not be difficult to show now that points P, K and Q are collinear and that angles $PQ\Omega$ and $QP\Omega$ are equal. Incidentally it appears that $C\Omega$ is an axis of symmetry for the parallel lines $A\Omega$ and $B\Omega$. Thus we are always able to construct a point Q on one of two parallel lines which corresponds to any point P on the other.

It is left to the reader to prove that there cannot be more than one point on $B\Omega$ which corresponds to a given point on $A\Omega$.

EXERCISE

Show that the line $C\Omega$ constructed above is unique for any two given parallel lines $A\Omega$ and $B\Omega$, and that it is the locus of all points equally distant from those lines. Thus, if $AB\Omega$ is regarded as a triangle with an ideal vertex, $C\Omega$ may be considered the bisector of the angle $A\Omega B$. It follows that the bisectors of the angles of such a triangle are concurrent.

Theorem 2. If three points P, Q and R lie one on each of three lines which are parallel to one another in the same sense, and if Q corresponds to P and R corresponds to Q, then the three points cannot be collinear.

For, if P, Q and R (Fig. 57) were collinear, the sum of angles $PQ\Omega$ and $RQ\Omega$ would equal two right angles and therefore so also would the sum of angles $QP\Omega$ and $QR\Omega$. But that would be absurd under the circumstances.

Theorem 3. If three points P, Q and R lie one on each of three lines which are parallel to one another in the same sense, and if Q corresponds to P and R corresponds to Q, then R corresponds to P.

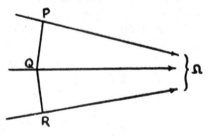

Figure 57

For the perpendicular to PQ (Fig. 57) at its midpoint is the axis of symmetry for the parallels $P\Omega$ and $Q\Omega$. Similarly the perpendicular to QR at its midpoint is the axis of symmetry for the parallels $Q\Omega$ and $R\Omega$. Since the perpendicular bisectors of two sides of the triangle PQR are parallel, that of the third side must be parallel to them in the same sense and it follows easily that P and R are corresponding points.

Corresponding Points on Intersecting Lines.

In the theory of corresponding points on intersecting lines, it is convenient to regard the intersecting lines as rays emanating from the point of intersection. With this viewpoint the student will be able to prove for corresponding points on intersecting lines the same three theorems proved above for corresponding points on parallel lines. The proofs are simpler than those for parallels.

Corresponding Points on Non-Intersecting Lines.

The reader will have no difficulty in proving, for corresponding points on non-intersecting lines, the three theorems proved above for corresponding points on parallel lines. In this case the lines under consideration have an ultra-ideal point in common, that is, they have a common perpendicular. There is one exception to Theorem 2 in this case.

58. Limiting Curves and Their Properties.

If, on any ray of a sheaf of rays with an ordinary point for vertex, a point is selected and then points corresponding to this one on other rays are constructed, these points will all lie on a circle. Indeed, a circle can be defined as the locus of a set of corresponding points on a sheaf of rays with an ordinary point for vertex. This alternative definition, from which can be derived all of the well-known properties of the circle, is the one referred to in Section 56. It is introduced here because it affords an easy transition from the familiar circle to new and strange types of curve. All that is necessary is to change the character of the sheaf of rays upon which the corresponding points lie.

In particular, let us consider the locus of a set of corresponding points on a sheaf of parallel lines, that is, on a sheaf with an ideal point for vertex. This locus is not a straight line in Hyperbolic

Geometry, for, by Theorem 2 of the preceding section, no three of its points are collinear. Nor is it a circle, although, owing to the similarity of its definition to that of the circle, it may be expected to have many properties in common with it. Since this curve is the limiting form approached by a circle as its radius becomes infinitely large, we shall call it a *limiting curve.*[16] The rays of the sheaf of parallel lines are called its *radii* or *axes*, and we shall occasionally refer to the ideal center of the sheaf as the *center* of the limiting curve.

Let us consider any two limiting curves PQ and $P'Q'$ with centers Ω and Ω', respectively (Fig. 58). Let A, B, C, D, etc., be any set of points chosen on PQ and draw the radii to them. On $P'Q'$ choose any point A' and draw the radius to it. Then construct angle $\Omega'A'B'$ equal to angle ΩAB and measure $A'B'$ equal to AB. If $B'\Omega'$

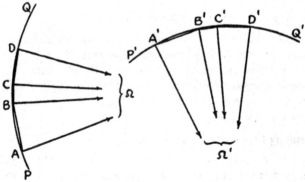

Figure 58

is drawn it is clear that angles $AB\Omega$ and $A'B'\Omega'$ are equal and hence also are angles $A'B'\Omega'$ and $B'A'\Omega'$. Therefore B' is on the limiting curve $P'Q'$. Continuing in this way, points C', D', etc., can be constructed on this limiting curve so that chords $B'C'$, $C'D'$, etc., are equal to chords BC, CD, etc., and the angles which pairs of corresponding chords make with the radii drawn to their extremities are equal also. We sum up these results by asserting that all limiting curves are congruent.[17] The correspondence just described can be

---

[16] Also frequently called a *horocycle.*

[17] See notes on modern ideas of congruence in Heath, *loc. cit.*, Vol. i, pp. 227–228.

set up on two arcs of limiting curves which are concentric or even on two arcs of the same limiting curve. We recognize the latter case by saying that a limiting curve has the same curvature[18] at all of its points.

Again, as a consequence of the investigations just outlined, we conclude that for any two limiting curves, or the same limiting curve, equal chords subtend equal arcs and equal arcs subtend equal chords. Here, of course, by equal arcs we mean arcs such as AD and $A'D'$ (Fig. 58) between the points of which there is a one-to-one correspondence of the kind described above. Furthermore, unequal chords subtend unequal arcs and the greater chord subtends the greater arc.

A straight line cannot cut a limiting curve in more than two points, for no three points of such a curve are collinear. If a line cuts a limiting curve in one point, and is not a radius, it will, in general, cut it in a second point. For example, let line AC (Fig. 59) cut

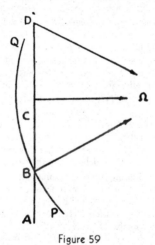

Figure 59

the limiting curve PQ in a point B. Draw the radius $B\Omega$. If we assume that angle $CB\Omega$ is acute, a distance corresponding to it as angle of parallelism can be constructed. Then, when BD is measured equal to twice this distance, D will fall on the limiting curve. Let the

[18] We use this word rather loosely here to convey the idea of the amount of bending.

line BD rotate about B. Point D will approach B as angle $DB\Omega$ increases and will coincide with B when angle $DB\Omega$ is a right angle. Thus a tangent line to a limiting curve at a point is perpendicular to the radius drawn to that point. In other words, a limiting curve cuts its axes at right angles and may be regarded as an orthogonal trajectory of its sheaf of radii. Furthermore, it is obvious that the curve is concave in the direction of parallelism of the radii.

Finally, it is not difficult to see that a line perpendicular to a chord of a limiting curve at its midpoint is a radius and that it bisects the arc subtended by the chord. Thus three points of a limiting curve determine it; when three of its points are given the center can be determined and other points constructed.

EXERCISES

1. How many limiting curves pass through two given points?

2. To draw a straight line in Hyperbolic Geometry a straight-edge is used; to draw a circle, compasses. What instrument is to be used in tracing limiting curves?

3. Prove that the segments of radii included between any pair of concentric limiting curves are equal.

4. Show that the radius drawn to the midpoint of an arc of a limiting curve bisects the corresponding arc of any concentric limiting curve, or, what is the same thing, that the line joining the midpoints of any two corresponding arcs of concentric limiting curves is a radius. Corresponding arcs are arcs included by any pair of common radii.

5. If points P_1, P_2, P_3, ... , P_{n-1} divide the arc AB of a limiting curve into n equal parts and the radii through these points cut the corresponding arc $A'B'$ of a concentric limiting curve in points P_1', P_2', P_3', ... , P'_{n-1}, then the latter points divide arc $A'B'$ into n equal parts.

6. Two corresponding arcs of two concentric limiting curves are unequal and the shorter arc lies in the direction of parallelism from the longer.

7. Given a point A on a limiting curve, the radius $A\Omega$, and nothing else. A line p is drawn perpendicular to $A\Omega$ at point C. Construct the two points in which p cuts the limiting curve. *Suggestion:*[19] Let B (Fig. 60) be one of the required points, construct right triangle ABC and use the conventional lettering. The existence of a right triangle with parts a, b, c, λ, μ assures the existence of a Lambert Quadrilateral with parts b, μ', l, a', c'. Measure CD on p equal to a' and draw the perpendicular DE from D to the tangent AE at point A of the limiting curve. Draw $B\Omega$ and $D\Omega$ and prove that the latter,[20] when produced in the opposite direction, is parallel to AE.

[19] Max Simon, *Nichteuklidische Geometrie in Elementarer Behandlung*, arranged and edited by K. Fladt (Leipzig and Berlin, 1925).
[20] The line $D\Omega$ is drawn here as though it were curved. The lines of Hyperbolic Geometry are as "straight" as those in Euclidean, but it is frequently convenient to represent them as curved when it is important to exhibit, within limited space, their asymptotic relationship to other lines, rather than their "straightness."

8. Given any point on a limiting curve, find a second point of the curve such that the tangent line at this point is parallel to the radius drawn to the first point. Do the same for a circle.

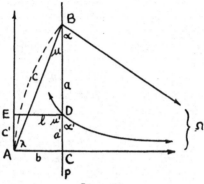

Figure 60

9. If a given point is so located that tangent lines can be drawn from it to a given limiting curve, show how to construct them. *Suggestion.*[21] Let P be the given point and draw $P\Omega$. It will cut the limiting curve in a point Q. Construct the tangent line to the limiting curve at Q. Then, by using Exercise 7, determine the points R and S in which the limiting curve concentric with the given one and passing through P is cut by this tangent. Let U and V be the points in which $R\Omega$ and $S\Omega$ cut the given limiting curve. PU and PV are the required tangents.

10. If a quadrilateral is inscribed in a limiting curve, prove that the sum of one pair of opposite angles is equal to the sum of the other pair.

59. Equidistant Curves and Their Properties.

Let us turn our attention next to the locus of a set of corresponding points on a sheaf of rays with an ultra-ideal vertex. This locus is not a straight line, nor is it a circle in the strict sense. From the theory developed for corresponding points on such a sheaf, we know that the curve is the locus of points which all have the same perpendicular distance from a straight line and are on the same side of it, this line being the representative line of the ultra-ideal point which is the vertex of the sheaf. For this reason it is called an *equidistant curve.* The representative line is referred to as its *base-line,* and the distance from every point on it to the base-line as the

---

[21] The construction described is essentially that given by Euclid for drawing the tangent from a point to a circle (III, 17). The fundamental character of his construction is indicated by the fact that it is independent of the parallel postulate and that it remains valid when the center of the circle becomes ideal.

distance. The rays of the sheaf are designated as the *radii* or *axes* of the equidistant curve and the ultra-ideal vertex may be regarded as its *center.* As a matter of fact, an equidistant curve properly consists of two branches, one on each side of the base-line. A straight line can be regarded as an equidistant curve with distance zero. In Euclidean Geometry the equidistant curve becomes a pair of parallel straight lines.

By methods similar to those used in the last section, the reader can readily substantiate the following statements:

Equidistant curves with equal distances are congruent, those with unequal distances are not. An equidistant curve has the same curvature at all of its points. Two curves with different distances have different curvatures, the greater the distance the greater the curvature. An equidistant curve is concave in the direction of the base-line.

For the same equidistant curve or congruent equidistant curves equal chords subtend equal arcs, and conversely. The reference is obviously to chords joining points of the same branch.

A straight line cannot cut an equidistant curve in more than two points. If a line cuts such a curve in one point it will cut it in a second unless it is tangent to the curve or parallel to its base-line. A tangent line to an equidistant curve is perpendicular to the radius drawn to that point, hence the curve may be regarded as an orthogonal trajectory of its sheaf of axes.

A line perpendicular to a chord of an equidistant curve at its midpoint is a radius and it bisects the arc subtended by the chord. Three points of such a curve determine it; when three of its points are given the base-line can be determined and other points constructed.

As a matter of interest, we call attention to the fact that three non-collinear points always lie on three different equidistant curves, since the vertices of a triangle are equidistant from each of the three lines joining the midpoints of the sides. We have already proved that the vertices of a triangle lie on a circle, using the word in the general sense; it will be a proper circle, limiting curve or equidistant curve according as the center is ordinary, ideal or ultra-ideal. In the latter case the three vertices lie on the same branch of the equidistant curve. Now we recognize that there are, in the general sense, four circles through the vertices of a triangle, just as there are four circles touching its sides.

EXERCISES

1. Show that the line joining any two points, one on each branch of an equidistant curve, is bisected by the base-line and makes with the branches, i.e., with their tangent lines, angles which are related to one another like those which a transversal makes with two parallels in Euclidean Geometry.

2. Let A and B be two points on one branch of an equidistant curve and A' and B' two points on the other so located that AA' and BB' intersect on the base-line. Construct lines AB' and BA' and compare the properties of the figure $ABA'B'$ with those of a parallelogram in Euclidean Geometry.

3. Given three points of an equidistant curve, construct other points (*a*) when the three points are on one branch, (*b*) when two lie on one branch and the third on the other.

4. The base-line for an equidistant curve and its distance are given. Construct the points in which a given line cuts the curve. There are three cases to be considered.

5. If a given point is so located that tangent lines can be drawn from it to a given equidistant curve, show how to construct them.

6. Devise an instrument to be used in tracing an equidistant curve when its base-line and distance are known.

7. Prove that, if a quadrilateral is inscribed in one branch of an equidistant curve, the sum of one pair of opposite angles is equal to the sum of the other pair. How must this statement be modified for a quadrilateral inscribed in an equidistant curve if all vertices do not lie on the same branch?

8. Carry out for an equidistant curve the instructions of problem 8, Section 58.

60. The Limiting Curve as Related to Circles and Equidistant Curves.

With the completion of our introduction to the elementary properties of circles, limiting curves and equidistant curves, it will perhaps be helpful to attempt to portray a little more clearly their relation to one another.

Let l and m (Fig. 61) be two perpendicular lines intersecting at O. Choose any point A on m, say to the right of l, and construct the circle with A as center and radius AO. If A is allowed to move toward O, the radius AO approaches zero and the circle, with its curvature increasing, approaches the point circle O as limiting form.

On the other hand, if A moves in the opposite direction, the circle increases in size, its curvature decreasing, and approaches as limiting form a limiting curve through O.

Next, choose any line n perpendicular to m at a point B and consider the equidistant curve with n as base-line and OB as distance. As n is moved toward l, say from the right, the curvature of the equidistant curve decreases and it approaches line l as a limiting form. However, as *n* moves away from *O*, the curvature increases and the equidistant curve approaches the limiting curve through *O* as limiting form.

On the left of *l* the situation is the same.

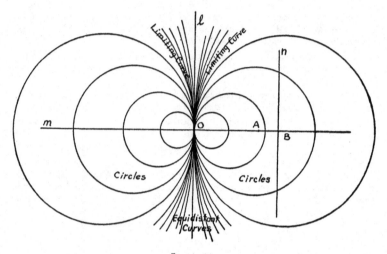

Figure 61

EXERCISES

1. Describe six properties common to all types of curve: circle, limiting curve, equidistant curve.

2. What procedure should be followed in attempting to determine whether three given points lie on a circle, limiting curve or one branch of an equidistant curve?

61. Area.

It will be remembered that, almost from the beginning of Book I, Euclid regards figures as *equal* when they are congruent. Not until I, 35 is reached do we find a modification of viewpoint. Here, without so much as calling attention to the change, he introduces the concept of figures which are equal but not congruent. The proposition referred to is the one which states that *parallelograms which are on the same base and in the same parallels are equal to one another*. It will be profitable to recall the proof.

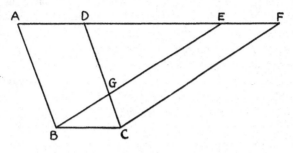

Figure 62

If *ABCD* and *EBCF* (Fig. 62) are the parallelograms, *BC* the
common base, and *AF* and *BC* the parallels, the student can easily
reproduce Euclid's proof by showing that triangles *EAB* and *FDC*
are equal, subtracting from each the triangle *DGE* and then adding
to each the triangle *GBC*. The proof is easily modified for the case
in which *D* and *E* coincide and the one in which *AD* and *EF* have a
segment in common.

Thus the equality of the two parallelograms results from the fact
that congruent figures have been subtracted from congruent figures
and then congruent figures added. Apparently nothing has been
used except the familiar common notions about equals added to and
subtracted from equals. It is to be observed, however, that the
assumption is tacitly made that it makes no difference *where* these
congruent figures are added or taken off.[22] It is helpful to recognize
this broader type of equality by the use of the word *equivalent*, as
did Legendre, who reserved the word *equal* for use in the sense of
congruent.

It should be made very clear that the proposition just stated and
proved is typical of all of those in the remainder of Book I and in
Book II which have to do with equal, or rather equivalent, figures.
These propositions are purely geometric; they are not metric in
character. No unit of area is used, nor indeed is the notion of area
introduced at all.

But from the idea of equivalence it is only a step to the concept of

[22] See Heath, *loc. cit.*, Vol. i, pp. 327–8.

area, a concept ordinarily associated with closed figures, just as the concept of length is associated with line segments. Areas or measures of area are thought of as magnitudes subject to the processes of addition and subtraction and the relations of equality and inequality. The theory of areas involves complications and difficulties. Even in Euclidean Geometry this is true, although to a degree matters are simplified there owing to the existence of the square. Attention is called to the fact that this use of a unit square for measuring areas implies an intimate relationship between the unit of area and the unit of length, a relationship which may occasionally be overlooked with resulting confusion.

In Hyperbolic Geometry there is no square. The quadrilateral with sides equal and angles equal has all of its angles acute. However, the general theory of equivalence and area has been placed upon a firm, logical foundation by modern investigators.[23] We propose merely to outline briefly for this geometry the theory as it appears in the light of recent developments.

62. Equivalence of Polygons and Triangles.

When two points on the perimeter of a polygon[24] are joined by a line segment or, more generally, by a broken line, all points of which lie within[25] the polygon, two new polygons are formed and the polygon is said to be *partitioned* into two polygons. If two polygons can be partitioned into the same finite number of triangles and a one-to-one correspondence established so that pairs of corresponding triangles are congruent, the two polygons are said to be *equivalent*. As a consequence of this definition the following theorem on the transmission of equivalence is easily proved.

Theorem 1. If two polygons are each equivalent to a third polygon, then they are equivalent to one another.

---

[23] See Hilbert, *loc. cit.*, pp. 69–82; the paper by Amaldi in Enriques' collection, *Questioni riguardanti la geometria elementare* (Bologna, 1900), or a translation into German, *Fragen der Elementargeometrie*, Vol. i, p. 151 (Leipzig, 1911); Max Simon, *Über die Entwicklung der Elementar-Geometrie im XIX. Jahrhundert*, pp. 115–121 (Leipzig, 1906); A. Finzel, *Die Lehre vom Flächeninhalt in der allgemeinen Geometrie* (Leipzig, 1912), or see *Mathematische Annalen*, 72 (1912), pp. 262–284.

[24] We refer here to *simple* polygons. See Section 9.

[25] See Section 9.

Let polygons P and Q (Fig. 63) each be equivalent to polygon R. Then to a partition of P into triangles corresponds a decomposition of R into an equal number of triangles congruent to them. Similarly to a partition of Q corresponds a partition of R.

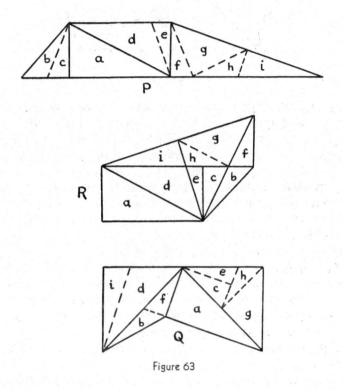

Figure 63

The lines for the two partitions of R divide it into triangles and polygons, and the polygons can, by the addition of segments if necessary, be partitioned into triangles. If, in the partitions of P and Q, lines are added corresponding to those inserted to complete the decomposition of R, it will appear that P and Q are thus divided into the same number of triangles, corresponding and congruent in pairs.

Theorem 2. Two triangles with a side of one equal to a side of the other and having the same defect are equivalent.

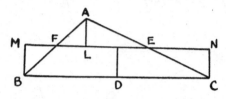

Figure 64

Let ABC (Fig. 64)[26] be any triangle with D, E and F the midpoints of sides BC, CA, and AB, respectively. Draw FE and construct the perpendiculars AL, BM and CN to it from the vertices. Then right triangles ALF and BMF are congruent and so also are right triangles ALE and CNE. As consequences, BM, AL and CN are equal, $BCNM$ is a Saccheri Quadrilateral which is equivalent to triangle ABC and has each summit angle equal to one-half the angle sum of triangle ABC, and FE is perpendicular to the perpendicular bisector of side BC of that triangle. The reader should verify these results for all positions of A with reference to B and C.

If one were to start with a second triangle $A'B'C'$, with side $B'C'$ equal to side BC of triangle ABC and with the same defect, it is clear that, by the same construction, it could be shown equivalent to a Saccheri Quadrilateral with the same summit and summit angles as the one obtained above and hence congruent to it. Since the two triangles are equivalent to congruent quadrilaterals they are equivalent to one another.

Theorem 3. Any two triangles with the same defect are equivalent.

[26] This beautiful construction and much of the accompanying argument is due to Henry Meikle (1844). See Frankland, *Theories of Parallelism*, p. 44 (Cambridge, 1910).

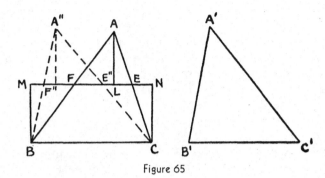

Figure 65

Let ABC and $A'B'C'$ (Fig. 65) be any two triangles with the same defect. It has already been proved that if a side of one is equal to a side of the other they are equivalent. Assume then that no side of one is equal to a side of the other, in particular that $A'C'$ is greater than AC.

As before, join F and E, the midpoints of sides AB and AC, and draw the perpendiculars AL, BM and CN from A, B and C to FE. Then locate on FE, on either side of N, a point E'' such that CE'' has length one-half that of $A'C'$. This can be done, since one-half of $A'C'$ is greater than CE, which in turn is greater than or equal to CN. Draw CE'' and produce it to A'' so that $E''A''$ is equal to CE''. Join A'' to B. Since FE is perpendicular to the perpendicular bisector of BC and cuts $A''C$ at its midpoint, it will cut $A''B$ at its midpoint F''. Then it is easy to show that triangles ABC and $A''BC$ have the same defect and are equivalent. But triangle $A''BC$ and $A'B'C'$ also have the same defect and a pair of equal sides and are therefore equivalent. Thus triangles ABC and $A'B'C'$ are equivalent to the same triangle and hence are equivalent to one another.

Theorem 4. Any two triangles which are equivalent have the same defect.

The proof of this theorem can be simplified by the use of the theory

of transversals as developed by Hilbert.[27]   A segment joining a
vertex of a triangle to a point on the opposite side is called a transversal.   A transversal divides a triangle into two subtriangles, and
one or both of these can be subdivided by transversals, and so on.
It is an easy matter to prove that if a triangle is subdivided by
arbitrary transversals into a finite number of subtriangles, the defect
of the triangle is equal to the sum of the defects of all of the triangles
in the partition.[28]   Furthermore, if a triangle is partitioned into
triangles in any way at all, this partition, if it is not already a
partition by transversals, can be made one by the addition of lines,
and as a consequence it will follow that the defect of the triangle will
be equal to the sum of the defects of all of the triangles in the
partition.   All that is necessary is to draw transversals from any
one of the vertices of the triangle through each vertex in the partition.   These transversals will divide the triangle into a number of
triangles with a common vertex.   Some or all of these are divided
by the lines of the partition into triangles and quadrilaterals, and
the quadrilaterals can be subdivided into triangles by the addition
of diagonals to complete the partition by transversals.

If two triangles are equivalent they can be subdivided into the
same finite number of triangles congruent in pairs.   Since the defect
of each triangle is equal to the sum of the defects of all of the triangles in the partition, it is clear that the two triangles will have the
same defect.

Finally, a triangle is said to be *equivalent to the sum of* two or more
triangles when the triangle can be partitioned into a finite number
of triangles and the two or more triangles can be partitioned altogether into the same number of triangles congruent to corresponding
triangles in the first partition.   Since the defect of a triangle is
equal to the sum of the defects of all of the triangles in any partition
of that triangle, we have the following theorem:

Theorem 5.   If a triangle is equivalent to the sum of two or more
triangles, its defect is equal to the sum of their defects.

---

[27] Hilbert, *Grundlagen der Geometrie*, 5th edition, pp. 57–60 (Leipzig and Berlin, 1922),
or the Townsend translation of the first edition, pp. 62–66.
[28] See Ex. 3, Section 44.

63. Measure of Area.

At this stage we change our viewpoint, with nice distinction. Since two triangles are equivalent when and only when they have the same defect, let us define the *measure of area* of a triangle, or simply the *area*, as the number obtained by multiplying the defect, expressed in terms of the unit described in Section 52, by a constant C^2. This constant is a proportionality factor which can be determined once a triangle is selected to have unity as its measure of area. It will play a prominent part in later developments and be subjected to some significant interpretations. The measure of area Δ of a triangle, with λ, μ and υ as the measures of its angles, is then given by the formula

$$\Delta = C^2(\pi - \lambda - \mu - \upsilon).$$

The student will now have no difficulty in verifying the following theorems:

Theorem 1. Two triangles have the same measure of area if and only if they are equivalent.

Theorem 2. If a triangle is partitioned into triangles in any way, the measure of area of the triangle is equal to the sum of the measures of area of all of the triangles in the partition.

Theorem 3. If a triangle is equivalent to the sum of two or more triangles, the measure of area of this triangle is equal to the sum of the measures of area of the two or more triangles.

Generalizing, we define the measure of area of a polygon as the sum of the measures of area of all of the triangles of any partition of the polygon into triangles. From what has already been said, it should be clear that this sum does not depend upon the particular partition used. If the difference between $(n - 2)$ straight angles and the sum of the angles of a polygon of n sides is called the *defect*[29] of the polygon, we see that the defect of a polygon is equal to the sum of the defects of all of the triangles in any partition. We conclude that the last three theorems can be generalized by replacing the word *triangle* by the word *polygon* wherever it occurs.

[29] See Ex. 4, Section 44.

64. The Triangle with Maximum Area.

In a letter[30] written in 1799 to the elder Bolyai, Gauss, who was at that time still trying to prove the Fifth Postulate, wrote:

"I have arrived at much which most people would consider sufficient for proof, but which proves nothing from my viewpoint. For example, if it could be proved that a rectilinear triangle is possible with an area exceeding any given area, I would be in a position to prove rigorously the whole of (Euclidean) geometry."

But he could not prove that there is no largest triangle. In fact, upon the assumption that there exists a triangle of maximum area, Gauss arrived at the formula, given in the preceding section, for the area of a triangle. This notable derivation, which he disclosed to Bolyai in 1832 in his letter[31] acknowledging receipt of the Appendix, is a beautiful bit of analysis. We present the essence of it here to conclude the discussion of area.

Gauss recognized, to begin with, that if there exists a triangle of maximum area it can be none other than that limiting form of triangle, with all of its vertices ideal points and its angles all zero angles, constructed by drawing a line parallel in opposite senses to two parallel lines. All such triangles are congruent and we shall assume that their common area is a constant δ.

Assume that the area enclosed by a straight line and the parallels to it through any point is a function of the angle between the parallels, say $f(\pi - \varphi)$. Then $f(\pi) = \delta$ and $f(0) = 0$. An examination of Figure 66(a) shows that

$$f(\pi - \varphi) + f(\varphi) = \delta,$$

while Figure 66(b) reveals that

$$f(\varphi) + f(\psi) + f(\pi - \varphi - \psi) = \delta.$$

Whence

$$f(\psi) + f(\pi - \varphi - \psi) = f(\pi - \varphi),$$

and we conclude that the function satisfies a functional equation of the form

$$f(\lambda) + f(\mu) = f(\lambda + \mu).$$

[30] See Section 30.
[31] See Section 30.

The solution[32] of this equation is

$$f(\theta) = c^2\theta,$$

where c^2 is a constant. Then

$$\delta = c^2\pi.$$

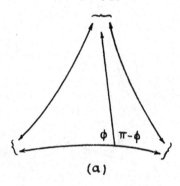

(a)

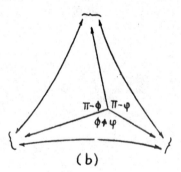

(b)

Figure 66

Turning now to any triangle, the angles of which measure λ, μ and v (Fig. 67), we produce its sides and draw common parallels to

---

[32] If the function is assumed to be continuous and to have a derivative, then it follows, since

$$\frac{f(\alpha + h) - f(\alpha)}{h} = \frac{f(\beta + h) - f(\beta)}{h},$$

that

$$f'(\alpha) = f'(\beta).$$

Thus

$$f'(\theta) = a$$

and

$$f(\theta) = a\theta + b,$$

where a and b are constants. But $b = 0$ since $f(0) = 0$.

them, taken in pairs. Then, designating the area of this triangle by Δ, we have

$$\Delta + f(\lambda) + f(\mu) + f(v) = c^2\pi$$

or finally

$$\Delta = c^2(\pi - \lambda - \mu - v).$$

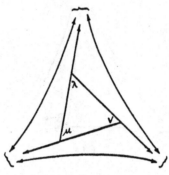

Figure 67

EXERCISES

1. Prove that the locus of the vertices of all triangles on the same base and having the same defect is an equidistant curve.

2. If E and F are the midpoints of sides AC and AB of triangle ABC, and E' and F' are the midpoints of sides $A'C'$ and $A'B'$ of triangle $A'B'C'$, prove that when EF and $E'F'$ are equal and the perpendiculars from A to EF and A' to $E'F'$ are also equal the triangles are equivalent.

3. On the hyperbolic plane, a regular network of regular polygons with n sides is to be constructed, p of them to meet at a point. Prove that the area of each polygon is

$$n\pi C^2\left(1 - \frac{2}{n} - \frac{2}{p}\right), \text{ with the condition } \frac{1}{n} + \frac{1}{p} < \frac{1}{2}.$$ Show also that the area of the

smallest finite regular quadrilateral with which the plane could be paved is $\frac{2}{5}\pi C^2$.

(Chrystal, 1880)

V

HYPERBOLIC PLANE TRIGONOMETRY

".... I can solve any problem in it with the exception of the determination of a constant, which cannot be designated *a priori*. The greater one takes this constant, the nearer one comes to Euclidean Geometry, and when it is chosen infinitely large, the two coincide." — Gauss

65. Introduction.

We turn next to an investigation of the trigonometry of the Hyperbolic Plane. In their developments of this theory, both Bolyai and Lobachewsky made use of the *limiting surface* or *horosphere*, the surface generated by the revolution of a limiting curve about a radius. It can be shown that upon that surface the geometry of the geodesics, which are limiting curves, is analogous to the geometry of straight lines on the Euclidean Plane.[1] But we shall derive the formulas for plane trigonometry without any appeal to solid geometry.[2]

66. The Ratio of Corresponding Arcs of Concentric Limiting Curves.

We begin by recalling certain relationships, already verified by the reader,[3] between *corresponding arcs* of concentric limiting curves, that is, between arcs included by a pair of common radii.

[1] See Sommerville, *The Elements of Non-Euclidean Geometry* (London, 1914), pp. 56, ff. and 84, for an elementary treatment from this standpoint.

[2] The development which we use follows closely that of Liebmann, *Nichteuklidische Geometrie*, 2nd edition, Chapter iii (Leipzig and Berlin, 1912). See also Gérard, *Sur la Géometrie Non Euclidienne*, Nouvelles Annales de Mathématiques, 1893, pp. 74–84. For a particularly careful treatment based upon the fact that Hyperbolic Geometry is Euclidean in character in an infinitesimal domain (Section 74), consult Coolidge, *Non-Euclidean Geometry*, p. 48, ff. (Oxford, 1909). For still another method with a different viewpoint, see W. H. Young, *On the Analytical Basis of Non-Euclidian* (sic) *Geometry*, American Journal of Mathematics, Vol. 33, 1911, pp. 249–286.

[3] See Section 58.

(*a*) Segments of radii included between any pair of concentric limiting curves are equal.

(*b*) The radius which bisects an arc of a limiting curve also bisects the corresponding arc of any concentric limiting curve.

(*c*) The lines joining the midpoints of any two corresponding arcs of concentric limiting curves is a radius.

(*d*) If points P_1, P_2, P_3, P_4, ... , P_{n-1} divide an arc AB of a limiting curve into n equal parts and the radii through the points of division cut the corresponding arc $A'B'$ of a concentric limiting curve in points P_1', P_2', P_3', P_4', ... , P'_{n-1}, then the latter points divide $A'B'$ into n equal parts.

Theorem 1. If A, B, C are three points on a limiting curve and A', B', C' are the points in which the radii through A, B, C cut a concentric limiting curve, then

$$\text{arc } AB : \text{arc } AC = \text{arc } A'B' : \text{arc } A'C'.$$

There are two cases to be considered.

First assume that arcs AB and AC (Fig. 68) are commensurable,

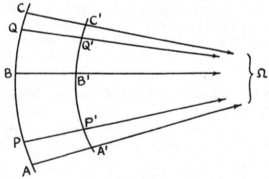

Figure 68

and let arc AP be a common unit of measure such that arc AB : arc $AP = m$ and arc AC : arc $AP = n$, m and n being integers. Draw the radius through P; it will cut $A'C'$ in a point P'. Then it is clear that arc $A'B'$: arc $A'P' = m$ and arc $A'C'$: arc $A'P' = n$ and thus that arc AB : arc AC = arc $A'B'$: arc $A'C'$.

Next assume that arcs AB and AC are incommensurable. If arc AP

is a unit of measure for arc AB, this unit can be applied to arc AC an integral number of times with arc QC as a remainder, where arc QC is less than arc AP. Draw the radius through Q; it will cut $A'C'$ in a point Q'. We have, from the case already proved,

$$\frac{\text{arc } AB}{\text{arc } AQ} = \frac{\text{arc } A'B'}{\text{arc } A'Q'}.$$

If the unit of measure, arc AP, is decreased, arcs QC and $Q'C'$ are decreased, and arcs AQ and $A'Q'$ approach arcs AC and $A'C'$, respectively, as limits. Then

$$\lim \frac{\text{arc } AB}{\text{arc } AQ} = \frac{\text{arc } AB}{\text{arc } AC}$$

and

$$\lim \frac{\text{arc } A'B'}{\text{arc } A'Q'} = \frac{\text{arc } A'B'}{\text{arc } A'C'}.$$

But if two variables are always equal and each approaches a limit, then the limits are equal. Therefore

arc AB : arc AC = arc $A'B'$: arc $A'C'$.

Next, on the line $P\Omega$ (Fig. 69), choose A, any point at all, and points B, C, D, E, ... , etc., so that $AB = BC = CD = DE = $ . . . . , etc. Let $Q\Omega$ be any line parallel to $P\Omega$ and points A_1, B_1, C_1, D_1, E_1, ... , etc., be the points on it corresponding[4] to A, B, C, D, E, ... , etc. Then $A_1B_1 = B_1C_1 = C_1D_1 = D_1E_1 = $ . . . . , etc. Draw the corresponding arcs of concentric limiting curves AA_1, BB_1,

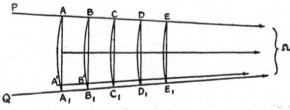

Figure 69

CC_1, DD_1, EE_1, etc. Since the arcs grow shorter in the direction of parallelism, arc AA' can be marked off on arc AA_1 equal to arc BB_1. If $A'\Omega$ is drawn, it will cut BB_1 in a point B'. As a consequence of

[4] See Section 57.

the congruence of figures $AA'B'B$ and BB_1C_1C, it follows that arcs BB' and CC_1 are equal and hence

$$\text{arc } AA_1 : \text{arc } BB_1 = \text{arc } BB_1 : \text{arc } CC_1 = \text{arc } CC_1 : \text{arc } DD_1 = \dots ,$$

etc.

Furthermore, if the distance AB between concentric arcs is increased, the ratio arc AA_1 : arc BB_1 is increased, and conversely. Thus we infer that the ratio of two corresponding concentric arcs depends not upon their position along the radii, nor yet upon the length of either arc, but only upon the distance between them.

Theorem 2. The ratio of two corresponding arcs of two concentric limiting curves depends only upon the distance between them, measured along a common radius.

At this point we are ready to select a unit of length. Since the ratio arc AA_1 : arc BB_1 (Fig. 69) is greater than unity, it can be made equal to e, the base of the natural system of logarithms, by proper choice of the length of AB. It is convenient to use AB, under these circumstances, as the unit of length. Thus again we recognize the absolute character of the unit of length in Hyperbolic Geometry. The particular unit just suggested is the one best adapted to the theoretical developments which are to follow. As a consequence of its transcendental character, it cannot, unfortunately, be constructed with straight edge and compasses.[5]

Thus, if we designate the arcs AA_1, BB_1, CC_1, etc., in Figure 69 by s, s_1, s_2, etc., and if $AB = BC = CD = \dots = 1$, we may write

$$\frac{s}{s_1} = \frac{s_1}{s_2} = \frac{s_2}{s_3} = \dots = \frac{s_{n-1}}{s_n} = e$$

and therefore

$$s_n = se^{-n},$$

where n is a positive integer.

It is now mere routine to prove the following theorem:

[5] See Liebmann, *Nichteuklidische Geometrie*, 2nd edition, §17 (Leipzig and Berlin, 1912).

Theorem 3. If s and s_x are the lengths of two corresponding arcs of concentric limiting curves, the direction of s_x from s being the direction of parallelism for the common radii, and if the radial distance between the arcs is x, then

$$s_x = se^{-x}. \tag{1}$$

Of course if x is irrational, that is, if the radial distance is incommensurable in terms of the unit of length, it will be necessary to use a limiting process such as the one used in the proof of Theorem 1.

A more general treatment results if the unit of length is so chosen that the ratio arc AA_1 : arc BB_1 (Fig. 69) is any constant a, greater than unity, when AB is the unit of length. Then we have

$$s_x = sa^{-x}.$$

Letting

$$a = e^{1/k},$$

we have

$$s_x = se^{-x/k}, \tag{2}$$

where, since $e^{1/k}$ is greater than unity, k is greater than zero. The number k is a parameter for Hyperbolic Geometry which depends for its value upon the choice of the unit of length, or, looked at from another angle, it is a constant, the choice of which determines the unit of length. It is the second constant which has been introduced.[6]

EXERCISES

1. Show that if s and s_x are the lengths of two corresponding arcs of concentric limiting curves, the direction of s_x from s being the direction of parallelism for the common radii, then the radial distance x between the arcs is given by

$$x = k \log \frac{s}{s_x}.$$

2. Given, in Euclidean Geometry, two intersecting lines PR and QR with distinct points A, B, C, D on PR such that $AB = CD$ and the corresponding[7] points A_1, B_1, C_1, D_1 on QR. Draw the arcs AA_1, BB_1, CC_1, DD_1 of concentric circles, center R. Prove that the ratio arc AA_1 : arc BB_1 cannot equal the ratio arc CC_1 : arc DD_1.

3. Derive Formula (2) by assuming that the ratio $s : s_x$ is a continuous function of x, say $f(x)$. Show first that $f(x) f(y) = f(x+y)$ and then, if $\log f(x) = F(x)$, that $F(x) + F(y) = F(x+y)$ and consequently $F(x) = hx$, where h is a constant.

[6] See Section 63.
[7] Section 57.

67. Relations between the Parts of an Important Figure.

We shall utilize the results of the preceding section to obtain some important formulas. But first it is to be observed that, given a point on a limiting curve, another point can be found on the curve such that the tangent line at the second point is parallel to the radius through the first. The length of this arc of limiting curve, such that the tangent at one extremity is parallel to the radius to the other, will be computed by the methods of calculus in Section 78, length of arc being defined in the usual way. We shall designate this length by S. It is obviously a constant, since a chord of length

$2p$ (Fig. 70), where $\Pi(p) = \dfrac{\pi}{4}$, subtends an arc of length $2S$. This

is the third constant which has thus far appeared. It will ultimately be revealed that they are all the same.

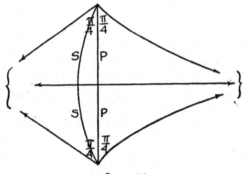

Figure 70

Let us start with an arc AB, of length s, of a limiting curve, center Ω (Fig. 71), such that s is less than S. Then if the tangent line is drawn to the curve at A, it will intersect the radius drawn to point B in a point C. Designate the length of AC by t, that of BC by u. By drawing chord AB, it is easy to show that in triangle ABC angle ABC is greater than angle CAB and hence t greater than u.

If arc AB is produced through B to D so that arc AD is equal to S, then the tangent line $AC\Omega'$ will be parallel to the radius through D. Next produce BC through C to a point E such that CE is equal to t. It follows that the perpendicular to CE at E will be parallel to $C\Omega'$

and consequently to $D\Omega'$. Construct the limiting curve, center Ω, through E and denote the point in which it intersects $D\Omega'$ by F. Arc EF is equal to S. The following relation is then easily obtained by the use of Theorem 3 of Section 66:

$$S - s = Se^{-(t+u)}. \tag{1}$$

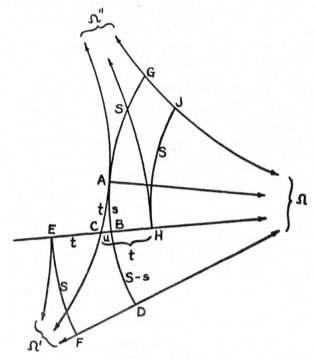

Figure 71

Now produce arc BA through A to point G such that arc AG is equal to S and draw the radius $\Omega G\Omega''$. This radius is parallel to the tangent line at A. Produce CB through B to a point H such that CH is equal to t and draw the limiting curve, center Ω, through H. This curve cuts ΩG in a point J. It is not difficult to show that the line through H parallel to $A\Omega''$ is perpendicular to CH and therefore is tangent to the limiting curve through H. We infer that arc HJ is equal to S and finally that

$$S + s = Se^{t-u}. \tag{2}$$

The addition of equations (1) and (2), member to member, gives[8]
$$e^u = \cosh t, \tag{3}$$
and subtraction yields, after a slight reduction, the result
$$s = S \tanh t. \tag{4}$$
Formulas (3) and (4) describe the relations between the parts of an important figure, the one composed of an arc of limiting curve of length less than S, the tangent at one extremity and the radius through the other. The student should be prepared to recognize this figure when it occurs and to recall these relations.

68. A Coördinate System and Another Important Figure.

At this stage it will be convenient to introduce a system of coördinates for the points of the Hyperbolic Plane. Let OX and OY (Fig. 72) be the familiar rectangular coördinate axes and P any

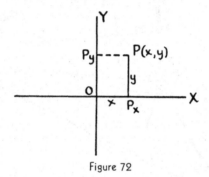

Figure 72

point. If P_x is the orthogonal projection of P on OX, then OP_x will be called the *abscissa* of P and P_xP the *ordinate*, with the usual convention for signs. It should be observed that, if P_y is the orthogonal projection of P on OY, the figure PP_xOP_y is a Lambert Quadrilateral and consequently that OP_y is not equal to y nor P_yP to x.

As an application of these ideas let us derive the equation of the limiting curve passing through the origin, with center the ideal point in the positive direction on the x-axis, and consequently with the y-axis as tangent at the origin. Choose on the curve (Fig. 73)

---

[8] The reader who is not acquainted with the *hyperbolic functions* and their properties will find a synopsis of the theory in the Appendix.

a representative point $P(x, y)$ with P_x its projection on the x-axis.

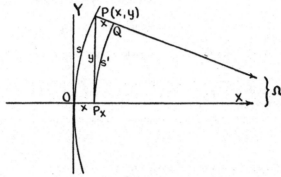

Figure 73

Let s represent the length of the arc OP. Designate by Q the point in which the concentric limiting curve through P_x cuts the radius through P and by s' the length of arc $P_x Q$. Then the equation of the limiting curve is readily obtained:

$$e^x = \cosh y. \tag{1}$$

Incidentally another important figure is encountered, namely, that formed by the arc of limiting curve of any length s, the perpendicular of length y from one extremity to the radius through the other and the segment x of this radius between the point in which it cuts the curve and the foot of the perpendicular. This is another figure to be recognized when it occurs. Since

$$s' = se^{-x}$$

and

$$s' = S \tanh y,$$

we obtain

$$s = S \sinh y. \tag{2}$$

Formulas (1) and (2) describe important relations between the parts of this figure.

In passing, we note that ΩP (Fig. 73) will intersect OY, be parallel to it or non-intersecting with respect to it, accordingly as s is less than, equal to or greater than S.

EXERCISES

1. What is the nature of the locus of the equation $y = c$, where c is a constant? of $x = c$?

2. Show that the equation of the limiting curve P_xQ (Fig. 73) is $e^{x-l} = \cosh y$, where $OP_x = l$.

3. If QP (Fig. 73) cuts the y-axis in a point R, prove that $\tanh b = \sinh y$, where $b = OR$.

4. Determine the equation of the straight line parallel to the x-axis with y-intercept equal to b. Consider the special case when b is infinite, that is, when the line is parallel to the y-axis also.

5. If s is the length of the arc of a limiting curve subtended by a chord of length l, show that $s = 2S \sinh \dfrac{l}{2}$.

69. The Relations between Complementary Segments.

We have already learned how, given a line segment, to construct the complementary segment. We are now in a position to find the analytical relation between any pair of complementary segments z and z'. Our objective can be easily attained if first we find the equation of the straight line parallel to each coördinate axis in the positive direction.

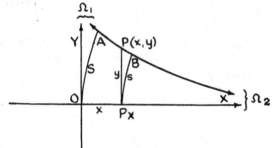

Figure 74

Let $\Omega_1\Omega_2$ (Fig. 74) be this line and $P(x, y)$ a representative point on it. Draw the arcs of concentric limiting curves OA and P_xB, center Ω_2, passing respectively through the origin O and the projection P_x of P on the x-axis, and included between that axis and the line $\Omega_1\Omega_2$. Designate by s the length of arc P_xB; the length of arc OA is S. Immediately we obtain the relations

$$s = S \tanh y$$

and

$$s = Se^{-x}.$$

Elimination of s yields the equation of $\Omega_1\Omega_2$,

$$e^{-z} = \tanh y. \tag{1}$$

To obtain the relation between any pair of complementary segments z and z', measure off from the origin in the positive direction on the x-axis (Fig. 75) segment OP_x equal in length to z. At P_x draw

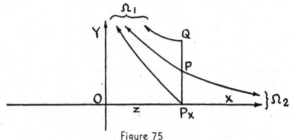

Figure 75

the perpendicular to the x-axis. It will cut $\Omega_1\Omega_2$, the common parallel to the coördinate axes, in a point P, as is easily shown by drawing $P_x\Omega_1$ through P_x parallel to the y-axis in the positive direction. Produce P_xP through P to Q so that PQ is equal to P_xP. At Q draw the line perpendicular to P_xQ. This perpendicular is parallel to $P\Omega_1$ and $P_x\Omega_1$. It is obvious that P_xQ is equal to z', the complementary segment of z. Since the coördinates of P are z and $\dfrac{z'}{2}$, we obtain from (1) the desired relation

$$e^{-z} = \tanh \frac{z'}{2}.$$

This result can be resolved into a more useful form as follows:
Since

$$\frac{e^z - e^{-z}}{2} = \frac{\coth \dfrac{z'}{2} - \tanh \dfrac{z'}{2}}{2}$$

$$= \frac{\cosh^2 \dfrac{z'}{2} - \sinh^2 \dfrac{z'}{2}}{2 \sinh \dfrac{z'}{2} \cosh \dfrac{z'}{2}}$$

$$= \frac{1}{\sinh z'},$$

we have

$$\sinh z = \operatorname{csch} z'.$$

It is important to be able to recognize this relation also in the forms

$$\cosh z = \coth z',$$
$$\tanh z = \operatorname{sech} z',$$
$$\operatorname{csch} z = \sinh z',$$
$$\operatorname{sech} z = \tanh z',$$
$$\coth z = \cosh z'.$$

70. Relations among the Parts of a Right Triangle.

In Euclidean Geometry we have an important relation, embodied in the remarkable Pythagorean Theorem, connecting the sides of a right triangle. There is also a useful relation between the acute angles. In Euclidean Trigonometry there are simple formulas relating the acute angles to pairs of sides in such a way that if two of these parts are given the third can be found. It is our next task to discover for Hyperbolic Geometry the analogous formulas relating the parts of a right triangle to one another by threes.

Let ABC (Fig. 76) be any right triangle and letter it in conven-

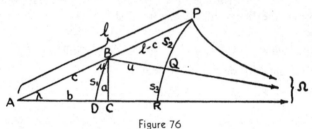

Figure 76

tional fashion. Measure off from A on the hypotenuse the segment AP equal to l, the segment which has λ for its angle of parallelism. Attention is called to the fact that in the drawing l has been assumed to be greater than c. This is not necessarily the case, but the reader can easily verify that the results will be the same when l is equal to or less than c. At P draw the line perpendicular to AP; it is parallel to $AC\Omega$. Draw also the line $B\Omega$ through B parallel to $AC\Omega$. Next construct two arcs of concentric limiting curves having $A\Omega$, $B\Omega$, $P\Omega$ as radii, namely, the one through B cutting $A\Omega$ in point

D and the one through P cutting $B\Omega$ in point Q and $A\Omega$ in point R. Designate the lengths of arcs BD, PQ, QR by s_1, s_2, s_3, respectively, and of segment BQ by u. The following relations are readily obtained:

$$s_1 = S \sinh a,$$
$$s_1 = s_3 e^u,$$
$$e^u = \cosh (l - c),$$
$$s_2 + s_3 = S \tanh l,$$
$$s_2 = S \tanh (l - c).$$

Then

$$\sinh a = \frac{s_1}{S} = \frac{s_3 e^u}{S} = e^u\left[\frac{s_2 + s_3}{S} - \frac{s_2}{S}\right]$$
$$= \cosh (l - c) [\tanh l - \tanh (l - c)]$$
$$= \frac{\sinh l \cosh (l - c) - \sinh (l - c) \cosh l}{\cosh l}$$
$$= \frac{\sinh c}{\cosh l}$$

or

$$\sinh c = \sinh a \cosh l. \tag{1a}$$

This is a relation connecting the hypotenuse, one leg and the opposite angle of the right triangle. The companion formula is

$$\sinh c = \sinh b \cosh m. \tag{1b}$$

It will be remembered that associated with every right triangle are four other right triangles, the existence of the first implying the existence of the others. We have learned that in order to recall this series of right triangles it is convenient to use a pentagon with sides lettered as directed in Section 53. If Formula ($1a$) is written

$$\cosh l = \sinh a' \sinh c,$$

it appears that the hyperbolic cosine of a middle part of the pentagon is equal to the product of the hyperbolic sines of the adjacent parts. By using each side in turn as the middle part, we pass from triangle to associated triangle and thus acquire other relations among the parts of the original right triangle. In this way, in addition to Formula ($1b$), we obtain

$$\cosh c = \sinh l \sinh m, \tag{2}$$

which is the formula relating the hypotenuse and two acute angles to one another, and

$$\cosh a' = \sinh b' \sinh l,$$

or

$$\tanh a = \frac{\sinh b}{\sinh l}, \tag{3a}$$

and its companion,

$$\tanh b = \frac{\sinh a}{\sinh m}, \tag{3b}$$

each of which provides a relation connecting the two legs with one of the acute angles.

If the values of $\sinh l$ and $\sinh m$, obtained from the last two formulas, are inserted in Formula (2), we obtain

$$\cosh c = \cosh a \cosh b. \tag{4}$$

This is the equivalent, in Hyperbolic Geometry, of the Pythagorean Theorem. When this result is interpreted in connection with the pentagon referred to above, we see that the hyperbolic cosine of a middle part is equal to the product of the hyperbolic cotangents of the opposite parts. By moving around the pentagon, we obtain, as results of the application of this rule, four more formulas:

$$\cosh m = \coth a' \coth l$$

or

$$\cosh a = \tanh l \cosh m, \tag{5a}$$

and

$$\cosh b = \tanh m \cosh l, \tag{5b}$$

each of which relates one leg to the acute angles, together with

$$\cosh a' = \coth c \coth m$$

and

$$\cosh b' = \coth c \coth l,$$

which are equivalent to

$$\tanh a = \tanh c \tanh m \tag{6a}$$

and

$$\tanh b = \tanh c \tanh l. \tag{6b}$$

Each of the latter two results connects the hypotenuse, a leg and the included angle.

All ten of these formulas, important in the study of the Hyperbolic Right Triangle from both the practical and theoretical viewpoints, can be written down easily by the use of the following rules suggested above and referring to the pentagon with sides lettered as in Section 53:

1. *The hyperbolic cosine of a middle part is equal to the product of the hyperbolic sines of the adjacent parts.*

2. *The hyperbolic cosine of a middle part is equal to the product of the hyperbolic cotangents of the opposite parts.*

The reader will immediately recognize the analogy between these rules and Napier's Rules for the spherical right triangle. These modifications of Napier's Rules are due to Engel and are known as the Napier-Engel Rules.

EXERCISES

1. Show that the equation of the line perpendicular to the y-axis, with y-intercept equal to b, is $\tanh y = \tanh b \cosh x$.

2. For the right triangle ABC, let the altitude CD from the vertex of the right angle to the hypotenuse have length h and divide the hypotenuse into the two segments $AD = p$ and $DB = q$. Prove that (a) $\tanh^2 a = \tanh c \tanh q$, (b) $\sinh^2 h = \tanh p \tanh q$.

71. Relations among the Parts of the General Triangle.

Directing our attention next to discovering formulas which relate the parts of triangles in general, we find ourselves in a position to derive the analogues of the Sine and Cosine Theorems, as we know them, in Euclidean Trigonometry. Just as those familiar formulas are usually obtained by dividing the triangle, through the construction of an altitude, into two right triangles, so are the corresponding formulas for the Hyperbolic Plane derived.

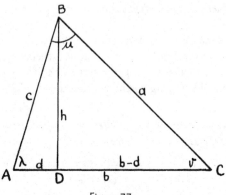

Figure 77

Let ABC (Fig. 77) be any triangle at all, λ, μ and υ its angles, a, b and c, respectively, the sides opposite these angles, and l, m and n the segments for which they are the angles of parallelism. Construct the altitude BD and designate its length by h. Then, applying Formula (1) of the preceding section to right triangles BDC and BDA, we obtain

$$\sinh a = \sinh h \cosh n,$$
$$\sinh c = \sinh h \cosh l,$$

and consequently

$$\frac{\sinh a}{\sinh c} = \frac{\cosh n}{\cosh l} = \frac{\text{sech } l}{\text{sech } n}.$$

By constructing a second altitude, we find that

$$\frac{\sinh b}{\sinh c} = \frac{\text{sech } m}{\text{sech } n},$$

so that

$$\sinh a : \sinh b : \sinh c = \text{sech } l : \text{sech } m : \text{sech } n. \qquad (1)$$

The reader will have no difficulty in proving that this result is valid even when an angle of the triangle is obtuse and an altitude must be drawn to a side produced, or when the triangle is a right triangle.

Returning to Figure 77, designate the length of AD by d and consequently that of DC by $b - d$. Then, by applying Formula (4) of the last section to the right triangles BDC and BDA, the following relations are secured:

$$\cosh a = \cosh b \cosh (b - d),$$
$$\cosh c = \cosh b \cosh d.$$

Then
$$\cosh a = \frac{\cosh c \cosh (b - d)}{\cosh d}$$
$$= \frac{\cosh c (\cosh b \cosh d - \sinh b \sinh d)}{\cosh d}$$
$$= \cosh b \cosh c - \sinh b \cosh c \tanh d.$$

This is easily reduced by means of the result obtained by applying Formula (6) of Section 70 to the right triangle BDA, namely,

$$\tanh d = \tanh c \tanh l.$$

Thus we obtain the analogue of the Cosine Formula, expressing the length of any side of a triangle in terms of the other two sides and the included angle:

$$\cosh a = \cosh b \cosh c - \sinh b \sinh c \tanh l. \tag{2}$$

Again the result will hold when one of the angles of the triangle is obtuse. When λ is a right angle, this relation resolves into Formula (4) of Section 70.

EXERCISES

1. If a transversal cuts the sides of triangle ABC, dividing side a into segments a_1 and a_2, side b into segments b_1 and b_2, and side c into c_1 and c_2 (Fig. 78), prove that

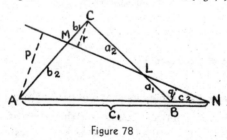

Figure 78

$$\sinh a_1 \sinh b_1 \sinh c_1 = \sinh a_2 \sinh b_2 \sinh c_2.$$

(The analogue of the Theorem of Menelaus.)

2. If the lines joining point P (Fig. 79) to the vertices of the triangle ABC divide

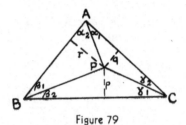

Figure 79

the angles at A, B and C, respectively, into the pairs of angles α_1 and α_2, β_1 and β_2, γ_1 and γ_2, prove that

$$\cosh a_1 \cosh b_1 \cosh c_1 = \cosh a_2 \cosh b_2 \cosh c_2,$$

where a_1 is the segment having α_1 as angle of parallelism, etc.

3. If the lines joining point P to the vertices of triangle ABC (Fig. 80) divide the

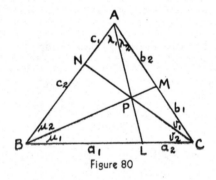

Figure 80

sides into pairs of segments a_1 and a_2, b_1 and b_2, c_1 and c_2, prove that

$$\sinh a_1 \sinh b_1 \sinh c_1 = \sinh a_2 \sinh b_2 \sinh c_2.$$

(Compare with the Theorem of Ceva.)

72. The Relation between a Segment and Its Angle of Parallelism.

We have already pointed out the fact that there is a functional relationship between a segment of line and its corresponding angle of parallelism. It is our next task to discover the formula which relates the two. If a is the length of any segment and α its angle of parallelism, we propose to prove that

$$\tanh a = \cos \alpha.$$

We begin[9] by examining the continuous function $f(\alpha)$, defined by the equation

$$\tanh a = \cos f(\alpha).$$

The following facts are readily verified:

When $\alpha = 0$, we have

$$a = \infty, \tanh a = 1, \cos f(\alpha) = 1, f(\alpha) = 0;$$

when $\alpha = \dfrac{\pi}{2}$, we find that

$$a = 0, \tanh a = 0, \cos f(\alpha) = 0, f(\alpha) = \frac{\pi}{2};$$

and when $\alpha = \pi$, it follows that

$$a = -\infty, \tanh a = -1, \cos f(\alpha) = -1, f(\alpha) = \pi.$$

Thus

$$f(0) = 0, f(\pi/2) = \pi/2, f(\pi) = \pi.$$

Consider any two angles λ_1 and λ_2. For simplicity we shall assume that each is acute and that their sum is acute. This restriction can be removed later without difficulty. Place these angles adjacent to one another (Fig. 81) and measure off on their common side, from

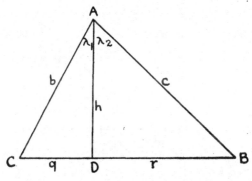

Figure 81

their common vertex A, a segment AD less than either of the segments corresponding to λ_1 and λ_2 regarded as angles of parallelism. Through D construct the line perpendicular to AD. It will cut the

[9] This treatment follows that of Liebmann, *Nichteuklidische Geometrie*, 2nd edition, pp. 75-77 (Leipzig and Berlin, 1912).

other sides of angles λ_1 and λ_2 in points C and B, respectively. Designate the segments AC, AB, CD, DB and AD by b, c, q, r and h, respectively.

Apply to the triangle ABC the Cosine Formula (Formula (2), Section 71) to obtain the relation:

$$\cosh (q + r) = \cosh b \cosh c - \sinh b \sinh c \tanh l,$$

where

$$\Pi(l) = \lambda_1 + \lambda_2.$$

Then

$$\cos f(\lambda_1 + \lambda_2) = \frac{\cosh b \cosh c - \cosh (q + r)}{\sinh b \sinh c}$$

$$= \coth b \coth c - \frac{\cosh q \cosh r}{\sinh b \sinh c} - \frac{\sinh q \sinh r}{\sinh b \sinh c}.$$

We shall reduce, in turn, the three terms on the right.

Observing, for right triangles ADC and ADB, that

$$\tanh h = \tanh b \tanh l_1$$

and

$$\tanh h = \tanh c \tanh l_2,$$

respectively, we conclude that

$$\coth b \coth c = \coth^2 h \cos f(\lambda_1) \cos f(\lambda_2).$$

For the second term, since

$$\frac{\cosh q}{\sinh b} = \frac{\cosh q \cosh h}{\sinh b \cosh h} = \frac{\cosh b}{\sinh b \cosh h}$$

$$= \frac{\tanh h}{\sinh b \tanh b} = \frac{\tanh l_1}{\sinh h},$$

and similarly

$$\frac{\cosh r}{\sinh c} = \frac{\tanh l_2}{\sinh h},$$

we have

$$\frac{\cosh q \cosh r}{\sinh b \sinh c} = \operatorname{csch}^2 h \cos f(\lambda_1) \cos f(\lambda_2).$$

Finally, since

$$\operatorname{sech} a = \sin f(\alpha),$$

it is easy to see that

$$\frac{\sinh q}{\sinh b} = \frac{\sinh q}{\sinh q \cosh l_1} = \operatorname{sech} l_1 = \sin f(\lambda_1)$$

and, in the same way,

$$\frac{\sinh r}{\sinh c} = \sin f(\lambda_2).$$

The net result is

$$\cos f(\lambda_1 + \lambda_2) = \cos f(\lambda_1) \cos f(\lambda_2) - \sin f(\lambda_1) \sin f(\lambda_2)$$
$$= \cos [f(\lambda_1) + f(\lambda_2)],$$

and we are led to the conclusion that the function under investigation satisfies the condition

$$f(\lambda_1) + f(\lambda_2) = f(\lambda_1 + \lambda_2).$$

Then we are able to write

$$\frac{f(\alpha_1 + h) - f(\alpha_1)}{h} = \frac{f(\alpha_2 + h) - f(\alpha_2)}{h}$$

and thus deduce, by allowing h to approach zero, that

$$f'(\alpha_1) = f'(\alpha_2)$$

and hence that $f'(\alpha)$ is constant. By integration we obtain the result

$$f(\alpha) = k\alpha + c,$$

where k and c are constants easily evaluated by taking into account the values of $f(\alpha)$ for α equal to o and $\frac{\pi}{2}$. And so it is manifest that

$$f(\alpha) = \alpha$$

and finally that

$$\tanh a = \cos \alpha. \tag{1}$$

This relation carries with it the following, as the reader can easily verify:

$$\coth a = \sec \alpha, \tag{2}$$
$$\operatorname{sech} a = \sin \alpha, \tag{3}$$
$$\cosh a = \csc \alpha, \tag{4}$$
$$\sinh a = \cot \alpha, \tag{5}$$
$$\operatorname{csch} a = \tan \alpha. \tag{6}$$

This important relationship connecting a segment and its angle of parallelism can be put into a somewhat more compact form by making use of the fact that

$$e^a = \sinh a + \cosh a.$$

152 NON-EUCLIDEAN GEOMETRY

We have at once that

$$e^a = \cot \alpha + \csc \alpha$$
$$= \frac{\cos \alpha + 1}{\sin \alpha} = \cot \frac{\alpha}{2}$$

and thus

$$\tan \frac{\alpha}{2} = e^{-a}, \tag{7}$$

a result obtained by both Bolyai and Lobachewsky.

73. Simplified Formulas for the Right Triangle and the General Triangle.

The important formulas obtained in Sections 70 and 71 can now be modified by use of the relations just obtained. Those for the right triangle become:

$$\sin \lambda = \frac{\sinh a}{\sinh c}, \tag{1a}$$

$$\sin \mu = \frac{\sinh b}{\sinh c}, \tag{1b}$$

$$\cot \lambda \cot \mu = \cosh c, \tag{2}$$

$$\tan \lambda = \frac{\tanh a}{\sinh b}, \tag{3a}$$

$$\tan \mu = \frac{\tanh b}{\sinh a}, \tag{3b}$$

$$\cosh c = \cosh a \cosh b, \tag{4}$$

$$\cosh a = \frac{\cos \lambda}{\sin \mu}, \tag{5a}$$

$$\cosh b = \frac{\cos \mu}{\sin \lambda}, \tag{5b}$$

$$\cos \mu = \frac{\tanh a}{\tanh c}, \tag{6a}$$

$$\cos \lambda = \frac{\tanh b}{\tanh c}. \tag{6b}$$

For the general triangle, we have

$$\sinh a : \sinh b : \sinh c = \sin \lambda : \sin \mu : \sin v$$

and

$$\cosh a = \cosh b \cosh c - \sinh b \sinh c \cos \lambda.$$

74. The Parameter.

In Section 66, we chose as our unit of length the radial distance between corresponding arcs of two concentric limiting curves, the ratio of the arcs being e. It was observed at the time that a more general treatment results if the unit is chosen as the radial distance when the two corresponding arcs have an arbitrarily chosen ratio a, where a is a constant greater than unity. In this way there is introduced into Hyperbolic Geometry a parameter k, greater than zero, such that

$$a = e^{1/k}.$$

The formula

$$s_x = se^{-x} \tag{1}$$

of Theorem 3, Section 66, then becomes

$$s_x = se^{-x/k}. \tag{2}$$

The preceding development made use of formula (1). If formula (2) is used, the results are the same, with the exception that in the formulas the lengths of all segments are divided by k. Thus the fundamental formula of Section 72 assumes the form

$$\tan \frac{\alpha}{2} = e^{-a/k},$$

while those of the last section become

$$\sin \lambda = \frac{\sinh a/k}{\sinh c/k}, \tag{1a}$$

$$\sin \mu = \frac{\sinh b/k}{\sinh c/k}, \tag{1b}$$

$$\cot \lambda \cot \mu = \cosh c/k, \tag{2}$$

$$\tan \lambda = \frac{\tanh a/k}{\sinh b/k}, \tag{3a}$$

$$\tan \mu = \frac{\tanh b/k}{\sinh a/k}, \tag{3b}$$

$$\cosh c/k = \cosh a/k \cosh b/k, \tag{4}$$

$$\cosh a/k = \frac{\cos \lambda}{\sin \mu}, \tag{5a}$$

$$\cosh b/k = \frac{\cos \mu}{\sin \lambda}, \tag{5b}$$

$$\cos \mu = \frac{\tanh a/k}{\tanh c/k}, \tag{6a}$$

$$\cos \lambda = \frac{\tanh b/k}{\tanh c/k}. \tag{6b}$$

The Sine and Cosine Formulas become, respectively,

$$\sinh \frac{a}{k} : \sinh \frac{b}{k} : \sinh \frac{c}{k} = \sin \lambda : \sin \mu : \sin v$$

and

$$\cosh \frac{a}{k} = \cosh \frac{b}{k} \cosh \frac{c}{k} - \sinh \frac{b}{k} \sinh \frac{c}{k} \cos \lambda.$$

If the parameter k is allowed to become infinite, it is significant that

$$\lim_{k \to \infty} \tan \frac{\alpha}{2} = \lim_{k \to \infty} e^{-a/k} = 1,$$

and the angle of parallelism for any distance approaches a right angle. We conclude that Hyperbolic Geometry becomes sensibly Euclidean if the parameter is chosen very large in comparison with the measures of the segments involved. As a matter of fact, all of the above formulas become those of Euclidean Geometry and Trigonometry as k tends to infinity.

For example, since

$$\lim_{k \to \infty} \frac{\sinh a/k}{a/k} = 1,$$

formula ($1a$) becomes

$$\sin \lambda = \frac{a}{c}.$$

Formula (4) becomes, upon replacing each hyperbolic cosine by the corresponding series and neglecting infinitesimals of higher order,

$$1 + \frac{c^2}{2k^2} = \left(1 + \frac{a^2}{2k^2}\right)\left(1 + \frac{b^2}{2k^2}\right),$$

or finally,

$$c^2 = a^2 + b^2.$$

The Sine and Cosine Formulas take, in the limit, the familiar forms

$$\sin \lambda : \sin \mu : \sin v = a : b : c$$

and

$$a^2 = b^2 + c^2 - 2bc \cos \lambda.$$

But there is another viewpoint which throws further light upon the situation. We can make $\frac{a}{k}$ infinitesimal by allowing a itself to approach zero, instead of letting k become infinite. Thus, for sufficiently small figures, we should expect the Euclidean formulas to hold approximately even in Hyperbolic Geometry. If careful experiments in the space in which we live, errors of measurement being taken into account, should seem to indicate that the sum of the angles of a triangle is always two right angles, that would not be convincing evidence that our space is strictly Euclidean in character. The sides of a triangle — even though the vertices are, for example, widely separated stars — may be too small in comparison with the parameter k and thus our space only apparently and approximately Euclidean.

That the formulas of Euclidean Geometry hold for the Hyperbolic Plane in the neighborhood of a point, that is, in an infinitesimal domain, is a fact of utmost importance. It affords the basis for the investigations to be made from the viewpoint of the calculus in the next chapter.

Finally, there is to be pointed out another fact which carries some significance. Any reader who is acquainted with spherical trigonometry has already recognized the resemblance which the formulas just derived for the right triangle and general triangle have to those for the triangles on a sphere. The constant k plays the role of the radius. Indeed, if k is imaginary the formulas are the same. Thus Hyperbolic Geometry may be regarded as analogous to geometry on a sphere with imaginary radius.[10]

EXERCISES

1. If s is equal to one-half the sum of the sides of the general triangle ABC, prove that

$$\cos\frac{\lambda}{2} = \sqrt{\frac{\sinh\frac{s}{k}\sinh\frac{s-a}{k}}{\sinh\frac{b}{k}\sinh\frac{c}{k}}}$$

and

$$\sin\frac{\lambda}{2} = \sqrt{\frac{\sinh\frac{s-b}{k}\sinh\frac{s-c}{k}}{\sinh\frac{b}{k}\sinh\frac{c}{k}}}.$$

2. Prove that the radius r of the inscribed circle of a triangle is given by

$$\tanh\frac{r}{k} = \sqrt{\frac{\sinh\frac{s-a}{k}\sinh\frac{s-b}{k}\sinh\frac{s-c}{k}}{\sinh\frac{s}{k}}}.$$

[10] See Section 23.

3. Obtain the limiting form, as k becomes infinite, for each of the formulas given in this section, connecting the parts of the right triangle and of the general triangle.

4. Obtain the limiting forms for the formulas of Exercises 1 and 3 of Section 71.

5. If the tangent at one extremity of an arc of a limiting curve of length s makes an angle θ with the subtending chord, prove that $s = 2S \tan \theta$. $(k = 1.)$

6. If the tangent at one extremity of an arc of length s of a limiting curve makes an angle φ with the radius through the other, show that $s = S \cos \varphi$.

7. Prove that the radius of the inscribed circle of a triangle of maximum area (Section 64) is $\frac{1}{2}k \log_e 3$.

8. If in a quadrilateral of maximum area, formed by drawing the four common parallels to two intersecting lines, the lengths of the common perpendiculars to the two pairs of opposite sides are a and b, prove that

$$\sinh \frac{a}{2k} \sinh \frac{b}{2k} = 1.$$

9. Three limiting curves are each tangent to all three sides of a triangle. Prove that the triangle is equilateral, that the measure of each side is $\cosh^{-1} \frac{3}{2}$ (using $k = 1$) and that each angle is $\cos^{-1} \frac{3}{5}$. Also show that the measure of the radius of the inscribed circle of the triangle is $\tanh^{-1} \frac{1}{4}$ and that of the radius of the circumscribed circle is $\tanh^{-1} \frac{1}{2}$. The three limiting curves here play the roles of the escribed circles of the triangle.

VI

APPLICATIONS OF CALCULUS TO THE SOLUTIONS OF SOME PROBLEMS IN HYPERBOLIC GEOMETRY

"It is upon the exactness with which we follow phenomena into the infinitely small that our knowledge of their causal relations essentially depends." — RIEMANN

75. Introduction.

There remain to be obtained a few results which are not to be omitted even from a brief introduction to the study of Hyperbolic Geometry. But such problems as the derivation of the formulas for the circumference and area of a circle call for the use of the calculus. Consequently, we direct our attention to finding the formulas for the differentials of arc and area in the geometry under consideration. These investigations are made comparatively simple by the fact discovered in the preceding chapter, namely, that the Euclidean formulas hold for infinitesimal triangles.

76. The Differential of Arc in Cartesian Coördinates.

Let the equation of a continuous curve in Cartesian coördinates be

$$y = f(x).$$

Let P (Fig. 82), with coördinates x and y, be any point on the curve and A a fixed point. Designate by s the length of the arc AP. Then s is a function of x. Following the usual procedure, we allow x to

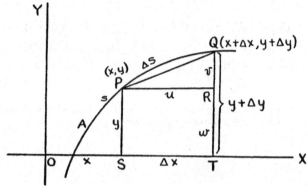

Figure 82

assume the increment Δx. Then y and s become $y + \Delta y$ and $s + \Delta s$, respectively. Denote by Q the point $(x + \Delta x, y + \Delta y)$. Next construct the perpendiculars PS and QT from P and Q to the x-axis, and draw PR perpendicular to QT. Designate the lengths of PR, QR and RT, respectively, by u, v and w. Then $PS = y$, $ST = \Delta x$ and $QT = v + w = y + \Delta y$. Finally allow Δx to approach zero. We have, for the infinitesimal right triangle PQR,

$$\overline{PQ}^2 = u^2 + v^2,$$

and thus

$$\frac{\overline{PQ}^2}{\Delta x^2} = \frac{u^2}{\Delta x^2} + \frac{v^2}{\Delta x^2}. \tag{1}$$

It is to be observed next that $PSTR$ is a Lambert Quadrilateral, acute-angled at P. The reader will readily recall[1] that this quadrilateral implies the existence of a right triangle with parts a, c, l and m', according to the conventional lettering, equal to Δx, u, y and w, respectively. Using relations connecting these parts of the right

[1] See Section 53.

triangle (Section 70, Formulas 1a and 5a), we have

$$\sinh \frac{u}{k} = \sinh \frac{\Delta x}{k} \cosh \frac{y}{k} \qquad (2)$$

and

$$\cosh \frac{\Delta x}{k} = \tanh \frac{y}{k} \coth \frac{w}{k}. \qquad (3)$$

From the first we observe that, except for infinitesimals of higher order,

$$u = \cosh \frac{y}{x} \Delta x,$$

while, for the second, it is clear that y and w differ by an infinitesimal, since the limit of the ratio of $\tanh \frac{y}{k}$ and $\tanh \frac{w}{k}$ is unity. We shall show that this infinitesimal is of higher order than Δx. Let

$$w = y - \epsilon.$$

Then from (3) we have

$$\tanh \frac{y}{k} = \tanh \frac{y - \epsilon}{k} \cosh \frac{\Delta x}{k},$$

which can be written

$$\tanh \frac{y}{k} = \frac{\tanh \frac{y}{k} - \tanh \frac{\epsilon}{k}}{1 - \tanh \frac{y}{k} \tanh \frac{\epsilon}{k}} \cosh \frac{\Delta x}{k},$$

or, except for infinitesimals of higher order,

$$\tanh \frac{y}{k}\left(1 - \frac{\epsilon}{k} \tanh \frac{y}{k}\right) = \left(\tanh \frac{y}{k} - \frac{\epsilon}{k}\right)\left(1 + \frac{\Delta x^2}{2k^2}\right).$$

Thus

$$\epsilon = \frac{1}{2k} \sinh \frac{y}{k} \cosh \frac{y}{k} \Delta x^2,$$

and ϵ is an infinitesimal of second order with respect to Δx. Furthermore, since

$$v = y + \Delta y - w = \Delta y + \epsilon,$$

it is clear that v and Δy are infinitesimals of the same order.

Returning to equation (1), assuming that arc Δs and chord PQ are equivalent infinitesimals, we may replace PQ, u and v by Δs, $\cosh \frac{y}{k} \Delta x$ and Δy, respectively, before taking limits to obtain the result

$$\left(\frac{ds}{dx}\right)^2 = \cosh^2 \frac{y}{k} + \left(\frac{dy}{dx}\right)^2,$$

and finally the formula for the differential of arc

$$ds^2 = \cosh^2\frac{y}{k}\,dx^2 + dy^2.$$

77. The Differential of Arc in Polar Coördinates.

Polar coördinates as used in Hyperbolic Geometry are defined in exactly the same way as in Parabolic Geometry Thus, in Figure 83,

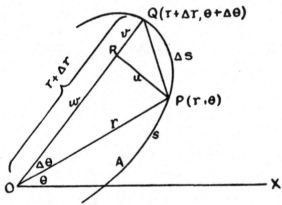

Figure 83

O is the pole and OX the polar line; the coördinates of P are r and θ, r being the radius vector and θ the vectorial angle. We shall obtain the formula for the differential of arc in polar coördinates.

Given the equation

$$r = f(\theta)$$

of a continuous curve in polar coördinates, let P (Fig. 83) be any point on the curve with coördinates r and θ, and A a fixed point. Denote by s the length of arc AP. When θ acquires the increment $\Delta\theta$, r becomes $r + \Delta r$ and s becomes $s + \Delta s$. Designate by Q the point $(r + \Delta r,\ \theta + \Delta\theta)$. Draw the radius vector OQ and chord PQ and also the perpendicular PR from P to OQ. Use the letters u, v and w to designate the segments PR, RQ and OR, respectively. Then, if $\Delta\theta$ is allowed to approach zero, we have for the right triangle PQR

$$\frac{\overline{PQ}^2}{\Delta\theta^2} = \frac{u^2}{\Delta\theta^2} + \frac{v^2}{\Delta\theta^2}. \qquad (1)$$

Using relations connecting the parts of right triangle ORP, we have (Section 74, Formulas 1 and 4)

$$\sin \Delta\theta = \frac{\sinh \frac{u}{k}}{\sinh \frac{r}{k}} \qquad (2)$$

and

$$\cosh \frac{r}{k} = \cosh \frac{u}{k} \cosh \frac{w}{k}. \qquad (3)$$

From the first it follows that u and $k \sinh \frac{r}{k} \Delta\theta$ are infinitesimals of the same order, while, from the second, we conclude that
$$r = w + \epsilon,$$
where ϵ is an infinitesimal. But ϵ is of higher order than $\Delta\theta$, since
$$\cosh \frac{w + \epsilon}{k} = \cosh \frac{w}{k} \cosh \frac{\epsilon}{k} + \sinh \frac{w}{k} \sinh \frac{\epsilon}{k},$$
and thus, neglecting infinitesimals of higher order,
$$\cosh \frac{w}{k} + \frac{\epsilon}{k} \sinh \frac{w}{k} = \left(1 + \frac{u^2}{2k^2}\right) \cosh \frac{w}{k},$$
so that
$$\epsilon = \frac{u^2}{2k} \coth \frac{w}{k}.$$

Then v and Δr are of the same order, since
$$v = r + \Delta r - w = \Delta r$$
to the lowest order.

Thus, in the limit, formula (1) becomes
$$ds^2 = k^2 \sinh^2 \frac{r}{k} d\theta^2 + dr^2.$$

78. The Circumference of a Circle and the Lengths of Arcs of Limiting Curve and Equidistant Curve.

We are now ready to apply the formulas of the last two sections to the solutions of some important problems. We obtain first the formula for the circumference of a circle, using the polar form of the equation, namely,
$$r = a.$$
Then

$$ds^2 = k^2 \sinh^2 \frac{r}{k} d\theta^2 + dr^2$$

$$= k^2 \sinh^2 \frac{a}{k} d\theta^2,$$

and

$$s = 4k \sinh \frac{a}{k} \int_0^{\pi/2} d\theta = 2k\pi \sinh \frac{a}{k}. \qquad (1)$$

We turn next to the equidistant curve, using Cartesian coördinates and the equation

$$y = b,$$

so that the x-axis is the base-line. Then, if s is the length of an arc of equidistant curve, the projection of which on the base-line is a, and if b is the common distance of all points from the base-line,

$$s = \cosh \frac{b}{k} \int_0^a dx = a \cosh \frac{b}{k}, \tag{2}$$

the length varying, as might have been predicted, directly as the projection.

We have already obtained, in Section 68, the equation in Cartesian coördinates of the limiting curve through the origin, with center the ideal point in the positive direction on the x-axis. In general form it is

$$e^{x/k} = \cosh \frac{y}{k}.$$

In order to find the length of an arc of this curve from the origin to any point (x, y), we note that

$$dx = \tanh \frac{y}{k} \, dy$$

and hence

$$s = \int_0^y \cosh \frac{y}{k} \, dy,$$

so that

$$s = k \sinh \frac{y}{k}. \tag{3}$$

The reader will be interested in determining the limiting forms of these results when k becomes infinite.

A particularly important discovery is made by obtaining the length S of an arc of limiting curve such that the tangent at one extremity is parallel to the radius through the other. All that is needed is to find the ordinate of the point of intersection of the limiting curve

$$e^{x/k} = \cosh \frac{y}{k}$$

and the line[2]

$$e^{-x/k} = \tanh \frac{y}{k},$$

[2] See Section 69.

parallel to both coördinate axes in the positive directions, and to use it in Formula (3). The ordinate is given by

$$y = k \sinh^{-1} 1,$$

and substitution yields[3]

$$S = k.$$

Thus it eventuates that two characteristic constants of Hyperbolic Geometry, introduced independently, are one and the same.

79. The Area of a Fundamental Figure.

Thus far, our investigations of area have been confined largely to that of the triangle. A very natural approach to the consideration of area from a more general viewpoint, and one which is quite in the manner of Euclid, is afforded by the ideas presented first in Section 66. It will be profitable and interesting to proceed a little way along this line of reasoning before taking up the study of area from the standpoint of the calculus and obtaining formulas for the element of area.

Let *ABC* (Fig. 84) be an arc of a limiting curve with center Ω,

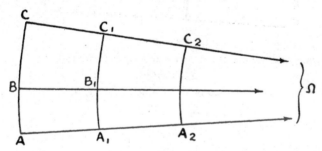

Figure 84

and A_1, B_1, C_1 the points in which the radii to *A*, *B* and *C*, respectively, cut a concentric limiting curve, which lies on the side of *ABC* in the direction of parallelism for the radii. Then, as a consequence of the results obtained in Section 66, and the fact that our concept of area includes the idea that congruent figures have equal areas, we conclude that

$$\frac{\text{Area } ABB_1A_1}{\text{Area } ACC_1A_1} = \frac{\text{arc } AB}{\text{arc } AC},$$

---

[3] This relation can also be obtained by comparing Formula (3) with Formula (2) of Section 68.

whether arcs AB and AC are commensurable with regard to one another or not. Furthermore, if the radii to A and C cut a third concentric limiting curve in points A_2 and C_2, respectively, this third curve lying on the side of A_1C_1 in the direction of parallelism for the radii, and if the radial distances AA_1 and A_1A_2 are equal, we have

$$\frac{\text{Area } ACC_1A_1}{\text{arc } AC} = \frac{\text{Area } A_1C_1C_2A_2}{\text{arc } A_1C_1}.$$

Thus it appears, for a figure comprising two corresponding arcs of concentric limiting curves and the segments of radii connecting their corresponding extremities, that the ratio of area to major arc depends not on the length of the arc, but on the radial distance between the arcs. Consequently we conclude, using the notation of Theorem 3, Section 66, that the area Δ_x (Fig. 85) included between

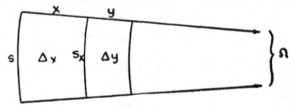

Figure 85

the corresponding concentric arcs s and s_x, is given by

$$\Delta_x = s\, f(x),$$

where $f(x)$ is a function to be determined. Then, if

$$\Delta_y = s_x\, f(y),$$

we have

$$\Delta_{x+y} = s\, f(x + y)$$

and thus

$$e^{x/k} f(x) + f(y) = e^{x/k} f(x + y).$$

In order to solve this functional equation, we interchange x and y, obtaining

$$f(x) + e^{y/k} f(y) = e^{y/k} f(x + y),$$

and then eliminate $f(x + y)$. The result is

$$\frac{f(x)}{1 - e^{-x/k}} = \frac{f(y)}{1 - e^{-y/k}},$$

so that

$$f(x) = C(1 - e^{-x/k}),$$

where C is a constant. The choice of C will determine the unit of area; it will be convenient to choose its value as k. Thus we finally have

$$\Delta_x = ks(1 - e^{-x/k}).$$

For the case in which s is equal to S, we see that

$$\Delta_x = k^2(1 - e^{-x/k})$$

and, as x becomes infinite, that the limit of this area is

$$\Delta_\infty = k^2.$$

This limit may be regarded as the area included by an arc of limiting curve of length S and the radii through its extremities. The unit of area has been so chosen that this area is k^2.

80. Limiting Curve Coördinates.

In order that we may take proper advantage of the results just obtained, we need to introduce a new system of coördinates peculiar to Hyperbolic Geometry.

As basis for reference in a system of *limiting curve coördinates*, we choose an axis $O\Omega$ (Fig. 86) — O being used as an origin, and Ω being

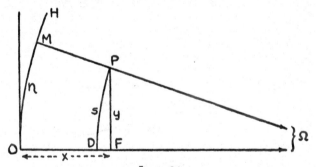

Figure 86

one of the ideal points on the axis — together with the limiting curve OH passing through O, and having Ω as center. Through any point P construct the limiting curve DP, concentric with the curve OH and cutting the axis at D, its radius $P\Omega$, when produced, cutting OH at M. Designate the lengths of OD and arc OM by ξ and η, respectively. These are the *limiting curve coördinates* of P.

We shall obtain the formulas for the transformation from limiting curve coördinates to Cartesian coördinates, when the axis and origin of the limiting curve coördinate system become the x-axis and origin of the Cartesian system. Then, if PF (Fig. 86) is drawn perpendicular to $O\Omega$, OF is the abscissa and PF the ordinate of the representative point P. Denoting arc PD by s, we know that

$$e^{(x-\xi)/k} = \cosh\frac{y}{k},$$

$$\eta = s e^{\xi/k},$$

$$s = S \sinh\frac{y}{k},$$

so that the desired equations are

$$\xi = x - k \log_e \cosh\frac{y}{k},$$

$$\eta = k e^{x/k} \tanh\frac{y}{k}. \tag{1}$$

81. The Element of Area.

The formula for the element of area is easily found in limiting curve coördinates by use of the ideas of area explained in Section 79. Let $P\,(\xi,\,\eta)$ (Fig. 87) be any point and Q a neighboring point with

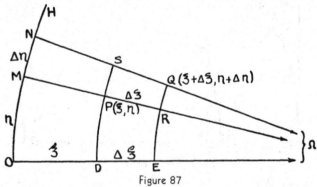

Figure 87

coördinates $\xi + \Delta\xi$ and $\eta + \Delta\eta$. Let the radius $P\Omega$ cut the limiting curves QE and OH in points R and M, respectively; let $Q\Omega$ cut the limiting curves PD and OH in points S and N. We choose the fundamental figure $PRQS$ as the element of area. Then, since

$$\text{arc } PS = \Delta\eta \cdot e^{-\xi/k},$$

we have

$$\text{area } PRQS = k \cdot \Delta\eta \cdot e^{-\xi/k}(1 - e^{-\Delta\xi/k}).$$

But, as $\Delta\xi$ and $\Delta\eta$ approach zero, $e^{-\Delta\xi/k}$ differs from $1 - \dfrac{\Delta\xi}{k}$ by infinitesimals of higher order, and the element of area becomes

$$e^{-\xi/k} \, d\xi \, d\eta. \tag{1}$$

The formula for the element of area in Cartesian coördinates can be obtained from this one by transforming it, using Formulas (1) of Section 80. From them come the total differentials

$$d\xi = dx - \tanh \frac{y}{k} \, dy, \tag{2}$$

$$d\eta = e^{x/k} \tanh \frac{y}{k} \, dx + e^{x/k} \operatorname{sech}^2 \frac{y}{k} \, dy. \tag{3}$$

But when (1) is used to obtain areas by integration, ξ must be regarded as constant while η is allowed to vary. Setting, under these circumstances, $d\xi = 0$ and eliminating dx, we have

$$d\eta = e^{x/k} \, dy,$$

so that (1) becomes

$$e^{\frac{x-\xi}{k}} \, d\xi \, dy. \tag{4}$$

Since, when ξ varies, y does not, we have from (2)

$$d\xi = dx,$$

and thus (4) finally becomes

$$\cosh \frac{y}{k} \, dx \, dy. \tag{5}$$

This formula for the element of area in Cartesian coördinates can also be obtained directly as follows:

Let P (Fig. 88) have coördinates x and y, and Q be the neighboring

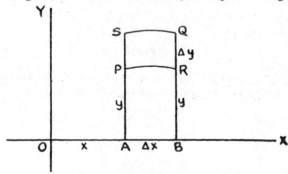

Figure 88

point with coördinates $x + \Delta x$ and $y + \Delta y$. Draw PA and QB perpendicular to OX. Let the equidistant curve through P, with OX as base-line, cut QB in R, and the one through Q cut PA in S. From Formula (2), Section 78,

$$\text{arc } PR = \cosh \frac{y}{k} \Delta x.$$

But, as Δx and Δy approach zero, the figure $PRQS$ becomes in the limit a rectangle. Thus for the element of area we have, as before,

$$\cosh \frac{y}{x} \, dx \, dy.$$

The element of area in polar coördinates can be obtained by transforming (5). However, we choose to obtain it directly from a figure.

Let P (Fig. 89) have coördinates r and θ, and Q be the point with

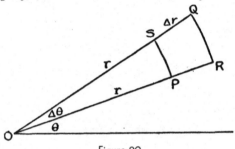

Figure 89

coördinates $r + \Delta r$ and $\theta + \Delta\theta$. Let the circle through P, with center at O, cut the radius vector through Q in point S, and let the circle through Q cut the radius vector through P in point R. From Formula (1), Section 78,

$$\text{arc } PS = k \sinh \frac{r}{k} \Delta\theta.$$

Since, as $\Delta\theta$ and Δr approach zero, the figure $PRQS$ becomes rectangular in form, we have for the element of area

$$k \sinh \frac{r}{k} \, dr \, d\theta. \qquad (6)$$

82. The Area of a Circle.

By making use of Formula (6) of the last section, we obtain the formula for the area of a circle. If the equation of the circle is taken as

$$r = a,$$

we have for the area

$$4k \int_0^a \int_0^{\pi/2} \sinh \frac{r}{k}\, d\theta\, dr,$$

which becomes

$$2\pi k^2 \left(\cosh \frac{a}{k} - 1 \right),$$

or, more compactly,

$$4\pi k^2 \sinh^2 \frac{a}{2k}.$$

It is to be observed that this area approaches πa^2, as k becomes infinite.

83. The Area of a Lambert Quadrilateral.

We next locate a Lambert Quadrilateral in a convenient position with reference to the coördinate axes and obtain its area.

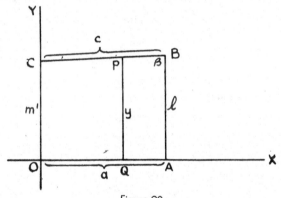

Figure 90

At any point A on the positive extension of the x-axis (Fig. 90), draw the perpendicular AB; from any point B on this perpendicular, draw BC perpendicular to the y-axis. Figure $OABC$ is then a representative Lambert Quadrilateral. We use the standard notation adopted in Section 53, designating segments OA, AB, BC and CO, respectively, by a, l, c and m', and angle ABC by β.

We shall need the equation of line CB. In order to obtain it, we locate upon CB a representative point P, with the perpendicular PQ as ordinate and OQ as abscissa. For the Lambert Quadrilateral $OQPC$, we have $OQ = x$, $QP = y$ and $OC = m'$. Using the relation between the corresponding parts of the associated right triangle, we get for the equation of CB

$$\tanh \frac{y}{k} = \frac{\cosh \dfrac{x}{k}}{\cosh \dfrac{m}{k}},$$

or, what is easily shown to be the same thing,

$$\sinh \frac{y}{k} = \frac{\cosh \dfrac{x}{k}}{\sqrt{\sinh^2 \dfrac{m}{k} - \sinh^2 \dfrac{x}{k}}}.$$

In order to find the area of quadrilateral $OABC$, we evaluate the integral

$$\int_o^a \int_o^y \cosh \frac{y}{k} \, dy \, dx,$$

obtaining, upon the first integration,

$$k \int_o^a \frac{\cosh \dfrac{x}{k}}{\sqrt{\sinh^2 \dfrac{m}{k} - \sinh^2 \dfrac{x}{k}}} \, dx,$$

and, upon the second,

$$k^2 \arcsin \frac{\sinh \dfrac{a}{k}}{\sinh \dfrac{m}{k}}.$$

But, from the results of Sections 70 and 72,

$$\frac{\sinh \dfrac{a}{k}}{\sinh \dfrac{m}{k}} = \tanh \frac{b}{k} = \cos \beta,$$

so that

$$\text{Area } OABC = k^2 \left(\frac{\pi}{2} - \beta \right).$$

84. The Area of a Triangle.

We are now but a step from the formula for the area of a triangle. The reader will recall how it was proved, in Section 62, that every triangle is equivalent to a Saccheri Quadrilateral for which the sum of the summit angles is equal to the angle sum of the triangle. But every Saccheri Quadrilateral can be divided into two congruent Lambert Quadrilaterals. Thus, if λ, μ and v are the angles of a triangle, its area is given by

$$2k^2 \left[\frac{\pi}{2} - \frac{\lambda + \mu + v}{2} \right],$$

or

$$k^2 \left[\pi - (\lambda + \mu + v) \right].$$

Reference to the formula for the area of a triangle, given in Section 63, shows that the constant C^2, introduced there, is equal to k^2.

EXERCISES

1. Show that the relation connecting the radius r of a circle, an arc of length s, and the angle θ, which the arc subtends at the center of the circle, is

$$s = k\theta \sinh \frac{r}{k}.$$

2. Show that the formulas for transformation from Cartesian coördinates to polar coördinates are

$$\tanh \frac{x}{k} = \tanh \frac{r}{k} \cos \theta,$$

$$\sinh \frac{y}{k} = \sinh \frac{r}{k} \sin \theta.$$

3. Use the transformation of Exercise 2 to change Formula (5) of Section 81 to Formula (6) of the same section.

4. Show that the equation of the equidistant curve $y = b$ is, in polar coördinates,

$$\sinh \frac{r}{k} \sin \theta = \sinh \frac{b}{k},$$

and, in limiting curve coördinates,

$$\eta = ke^{\xi/k} \sinh \frac{b}{k}.$$

5. Obtain the area bounded by the equidistant curve $y = b$, the x-axis and the ordinates $x = 0$ and $x = a$.   Ans. $ka \sinh \frac{b}{k}$.

6. Obtain, by integration, the area of the segment bounded by a chord of length $2l$ of a limiting curve and the subtended arc. Hint: Use the equation $e^{x/k} = \cosh \frac{y}{k}$, and

find the area bounded by this curve, the x-axis and the ordinate $x = k \log \cosh \dfrac{l}{k}$.

$$\text{Ans.} \quad 2k^2 \left(\sinh \frac{l}{k} - \arctan \sinh \frac{l}{k} \right).$$

7. Find, by integration, the area bounded by the x- and y-axes and the line $e^{x/k} = \dfrac{\tanh l/k}{\tanh y/k}$, parallel to the x-axis, and having its y-intercept equal to l.

$$\text{Ans.} \quad k^2 \left(\frac{\pi}{2} - \lambda \right), \text{ where } \lambda = \Pi(l).$$

8. By combining half the area of Exercise 6 with the area of Exercise 7, obtain the formula for the area bounded by an arc of limiting curve of length s and the radii to its extremities.

$$\text{Ans.} \quad k^2 \sinh \frac{l}{k} = ks.$$

VII

ELLIPTIC PLANE GEOMETRY AND TRIGONOMETRY

> "The unboundedness of space possesses a greater empirical certainty than any external experience. But its infinite extent by no means follows from this." — RIEMANN

85. Introduction.

The charácteristic postulate of Euclidean Geometry states that, through a given point, one and only one line can be drawn which is parallel to a given line. On the other hand, the distinctive feature of Hyperbolic Plane Geometry is the assumption that an infinite number of parallels can be drawn to a line through a point. It is now incumbent upon us to investigate, although briefly, the consequences of a third supposition, namely, that *no* line can be drawn through a given point, parallel to a given line. This we recognize as equivalent to the hypothesis of the obtuse angle of Saccheri. He and others were able to rule out the geometry based upon it, because they expressly or tacitly assumed that straight lines are infinite. It will be recalled that we have already shown[1] that these two assumptions are incompatible. To make this clearer, we remark that, if straight lines are infinite, then the proof of Euclid I, 16 is valid and consequently that of I, 27. But in this case there is always one parallel, at least, to a line through an exterior point.

It was Riemann[2] who first pointed out the importance of distin-

---

[1] See Chapter II.
[2] See Section 6.

173

guishing between the ideas of *unboundedness* and *infinitude* in connection with concepts of space. However strongly convinced we may be of the endlessness of straight lines, it does not necessarily follow that they are infinite in extent.

Therefore, before formally stating the characteristic postulate of Elliptic Geometry, we replace Euclid's tacit assumption of the infinitude of the line by a milder one:

POSTULATE. *Every straight line is boundless.*

The characteristic postulate of Hyperbolic Geometry is compatible with all of the postulates of Euclidean Geometry except the one which it replaces. Indeed, it was the similarity of those two geometries in their foundations and early propositions which enabled us to present an account of Hyperbolic Geometry without long and confusing preliminaries.

But the transition from Euclidean Geometry to Elliptic is not so easily accomplished. The characteristic postulate of Elliptic Geometry, stated in the next section, is not only incompatible with the Euclidean postulate which it replaces, and with the one which asserts that straight lines are infinite, but with others, as we shall see. Furthermore, it should be observed critically that those propositions in Euclidean Geometry which depend explicitly on the infinitude of the line — in particular I, 16 and its consequences — will no longer be valid in general. A more extensive account than we propose to give here would require a very carefully laid foundation.

86. The Characteristic Postulate of Elliptic Geometry and Its Immediate Consequences.

We are now in a position, having made the above change, to introduce the *Characteristic Postulate* of Elliptic Plane Geometry.

POSTULATE. *Two straight lines always intersect one another.*

Let l (Fig. 91) be any line at all. At any two points A and B on this line draw the lines[3] perpendicular to it. These, as a conse-

---

[3] Some of the lines in this figure and subsequent ones are drawn as though they were curved. The lines of Elliptic Geometry are as "straight" as those of Euclidean or Hyperbolic Geometry. It is frequently convenient to represent them as curved when it is of more importance to show, within limited space, their relation to each other than to exhibit their "straightness."

quence of the characteristic postulate, will meet at a point O and, since the angles at A and B in triangle AOB are equal, it follows that OA and OB are equal.[4] If AB is produced in either direction, say

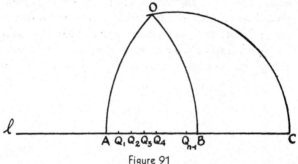

Figure 91

through B, to C, so that BC is equal to AB, and if OC is drawn, then it is easy to show that OC is perpendicular to l and is equal to OA and OB.[5] By repeating this construction, we are led to conclude that, given a segment AB of a line l, if P is any point on l such that AP equals $m \cdot AB$, where m is a positive integer, then the perpendicular to l at P passes through O, the intersection of the perpendiculars to l at A and B, and OP is equal to OA.

Next divide AB into n equal parts, $Q_1, Q_2, Q_3, \ldots, Q_{n-1}$ being the points of division. The perpendicular to l at Q_1 will cut AO at O, for, if it were to meet it at another point, so also — as is clear from what has already been proved — would the perpendicular at B. The same is of course true for the perpendiculars at the other points of division. Reasoning in this way, we conclude that, if segments AB and AP are commensurable, the perpendicular at P will pass through O and OP will equal OA. When AB and AP are incommensurable, the same results are obtained by limiting processes in the usual way.

Thus the perpendiculars at all points of a line are concurrent in a point called the *pole* of the line. Every line joining a point of a line

---

[4] It will appear presently that A and B may, by chance, be so situated that the two perpendiculars will be the same line. This complication can be avoided by changing the position of one of the points. The proof of Euclid I, 6 holds here if A, O and B are not collinear.

[5] The proof of Euclid I, 4 holds for Elliptic Geometry. See Section 5.

to its pole, or, what is the same thing, every ray emanating from the pole of a line, is perpendicular to the line. The reader will have no difficulty in showing that not only is the perpendicular distance from the pole to the line the same regardless of which perpendicular is used, but that the distance from pole to line is the same for all lines. Let us designate this constant perpendicular distance by q.

Continuing our investigations, let O (Fig. 92) be the pole of the line l. Draw any two lines through O; they will cut l at right

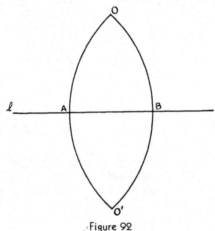

Figure 92

angles in two points A and B. Produce OA through A to O', so that AO' is equal to q. Then if O' and B are joined, it is easy to show that $O'B$ is perpendicular to l, that O, B and O' are collinear, and that BO' is of length q. Thus, rejecting for the moment the possibility of conceiving that O and O' are the same point, it appears that every line has *two* poles. Furthermore, lines OA and OB have a common perpendicular and intersect in *two* points, forming a *digon*, or *bi-angle*, each side of which is of length $2q$. This is true of every pair of lines, as we shall now show.

Let l and m (Fig. 93) be any two lines. They will intersect at some point O. Measure off from O on each line, and in each direction, a segment equal to q. In particular, let OA, OB, OC and OD be of length q. Then A, B, C and D will all lie on a line n for which O is a pole.[6] As a consequence, l and m intersect in another point O',

[6] See Exercise 2, Section 90.

the second pole of *n*. It follows that two lines in Elliptic Geometry always have a unique common perpendicular and always intersect one another in two points, enclosing an area. Furthermore, it is now apparent that each line returns into itself, i.e., is closed or

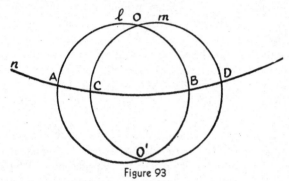

Figure 93

re-entrant, and is thus finite and of length $4q$. Two points do not always determine a line, for if the two points are the poles of a line, an infinite number of lines can be drawn through them. It should be recognized here that while, as a consequence of the finitude of the line, Euclid I, 16 and the propositions dependent upon it are not valid, in general, for Elliptic Geometry, nevertheless, they will continue to hold if the figures involved are small enough. For example, if each median of a triangle has a length less than *q*, then each exterior angle is greater than either of the opposite, interior angles. For figures in Elliptic Geometry which are sufficiently restricted in size, Euclid I, 17, 18, 19, 20 and 21 will continue to hold.

87. The Relation between Geometry on a Sphere and Elliptic Geometry.

Under the assumption made above, that a line has two distinct poles, there arises a geometry like that on a sphere, if great circles are regarded as representing straight lines. As far as we have gone the analogy is easily traced. For example, two great circles on a sphere always intersect one another in two points and enclose an area; every great circle has two poles through which pass all of the great circles orthogonal to it; two points determine a great circle, unless they are the poles of a great circle; the great circles of a sphere are finite, closed, and all of the same length.

It must not be inferred at all, however, that this type of Elliptic Geometry *is* spherical geometry. It merely happens that, on the curved surface known as a sphere, we find an exact representation of this kind of plane geometry, entity for entity, postulate for postulate, proposition for proposition. The reader will understand the significance of this relation better, if he is informed that there are curved surfaces in Hyperbolic Geometry and Elliptic Geometry upon which analogues of Euclidean Plane Geometry can be constructed.[7] It should be noted, in passing, that spherical geometry itself is entirely independent of the postulate on parallels.

In any attempt to visualize Elliptic Geometry, this resemblance to spherical geometry will be found quite helpful. Comparison makes it easier to understand how the sum of two sides of a triangle can be less than the third, how a triangle with a pair of equal angles may have the sides opposite them unequal, how the greatest side of a triangle may not subtend the greatest angle, how a Saccheri Quadrilateral can have its summit angles larger than two right angles, why even Pasch's Axiom will not always hold.

This is perhaps an appropriate place to remark that there also exist curved surfaces in Euclidean Space upon which can be constructed representations of Hyperbolic Plane Geometry. The fact that the sphere of radius r is a surface of constant positive curvature $\dfrac{1}{r}$ suggests[8] that for this purpose one should seek a real surface of constant negative curvature. An example of such a surface is that obtained by revolving the curve known as the *tractrix* about its asymptote. The equation of the tractrix is

$$x = a \log \frac{a + \sqrt{a^2 - y^2}}{y} - \sqrt{a^2 - y^2}.$$

This surface, called the *pseudosphere*, has a constant total or *Gaussian* curvature $-\dfrac{1}{a^2}$, and is one upon which, with restrictions, a geometry analogous to Hyperbolic Plane Geometry can be constructed, the geodesics playing the roles of straight lines. But further investi-

---

[7] See Section 65.   [8] See Sections 23 and 32.

gations in this direction require the employment of the methods of differential geometry[9] and carry us beyond the scope of this book.

88. The Two Elliptic Geometries.

Thus far we have proceeded on the assumption that the two points O and O' of Section 86 are distinct. However, it is conceivable that they are the same point and that a line has only one pole. This viewpoint leads to a perfectly consistent geometry.

If a line is regarded as having but one pole, two lines intersect in only one point and two points always determine a line. Straight lines are finite and closed, but of length $2q$. The distinguishing feature of this type of Elliptic Geometry is the fact that a straight line

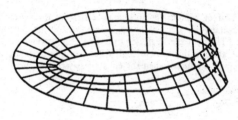

Figure 94

does not divide the plane into two regions. In other words, it is always possible, on such a plane, to pass from one side of a line to the other without crossing the line.

The distinction between the two Elliptic Planes can, perhaps, be made clearer by calling attention to the fact that the plane in the geometry sometimes described as of the *spherical* type has the character of a *two-sided* surface. On the other hand, the plane in the Elliptic Geometry with a single pole for each line is *one-sided* in character. One usually conceives the Euclidean Plane, for example, as having two faces, and a sphere as having two distinguishable surfaces, referred to, very likely, as *inside* and *outside*. The concept of a strictly one-sided surface is less familiar. It will aid in formulating our ideas of such a surface to consider what is called a *Leaf* (or *Sheet*) of *Möbius* (Fig. 94). This can easily be constructed by twisting, halfway around, a long, narrow, rectangular strip of paper, and then

[9] See, for example, Eisenhart: *Differential Geometry*, Chapter VIII (Boston, 1909), or Graustein: *Differential Geometry*, p. 179 (New York, 1935).

pasting the ends together. Thus the two faces of the rectangle become indistinguishable and the resulting surface has only one side. The drawing suggests how it is possible to pass from one side of a line to the other, so to speak, without crossing the line.

These two types of Elliptic Plane are generally distinguished by designating one as the *Double Elliptic Plane* and the other as the *Single Elliptic Plane.* As remarked previously,[10] the first — the spherical type — is probably the one which Riemann had in mind.

The following brief treatment of Elliptic Geometry and Trigonometry is confined chiefly to that portion of the theory common to both planes.

EXERCISE

Construct a Leaf of Möbius, but first draw a straight line lengthwise down the middle of the rectangular strip of paper, before pasting the ends together as instructed above. Then cut the Leaf along the line. Interpret the result. Repeat, this time cutting along a line drawn lengthwise, but only one-third of the distance across the strip, and produced when the ends are pasted together.

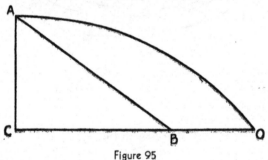

Figure 95

89. Properties of Certain Quadrilaterals.

In the subsequent developments we shall need the following lemma:

Lemma. In any triangle which has one of its angles a right angle, each of the other two angles is less than, equal to, or greater than a right angle, according as the side opposite it is less than, equal to, or greater than q, and conversely.

Let angle C in triangle ABC (Fig. 95) be a right angle. Measure off from point C on side CB, in the direction of B, a segment CO

[10] See Section 35.

equal to q. Line AC then has point O for pole. If O is joined to A, angle OAC is obviously a right angle. Hence, if CB is less than q, as in the drawing, angle CAB will be less than a right angle, while, if CB is equal to or greater than q, that angle will be equal to or greater than a right angle. The converse is then easily proved.

Theorem 1. The line joining the midpoints of the base and summit of a Saccheri Quadrilateral is perpendicular to both of them, and the summit angles are equal and obtuse.

It will only be necessary for us to show here that the summit angles are obtuse. The proof of the remainder of the theorem is the same as that given for the corresponding theorem in Hyperbolic Geometry.

Let $ABCD$ (Fig. 96) be a Saccheri Quadrilateral with base AB and with MH the line joining midpoints of base and summit. We need only consider cases in which each summit angle is known to be less than two right angles. Produce HC and MB until they meet in a

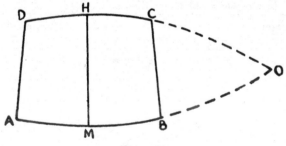

Figure 96

point O. The line HM then has O for pole and segments HO and MO are of length q. Then the length of BO is less than q and angle BCO is acute. It follows that the summit angles are obtuse.

Theorem 2. In a trirectangular quadrilateral (Lambert Quadrilateral) the fourth angle is obtuse and each side adjacent to this angle is smaller than the side opposite.

The proof that the fourth angle is obtuse is left to the reader. To show that the sides of the obtuse angle are less than their respective opposite sides, let *ABCD* (Fig. 97) be a Lambert Quadrilateral with obtuse angle at *C*. Then *BC* cannot equal *AD*, for then the angles at

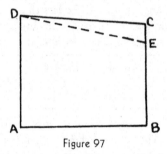

Figure 97

D and *C* would have to be equal. Nor can *BC* be greater than *AD*. For, if such is the case, measure off segment *BE* on *BC* equal to *AD*. Then angles *ADE* and *BED* will be equal. But angle *ADE* is acute and another contradiction has been reached. We conclude that *BC* is less than *AD*.

90. The Sum of the Angles of a Triangle.

Theorem 1. If any triangle has one of its angles a right angle, then the angle-sum of the triangle is greater than two right angles.

One needs to consider only cases in which each of the other two angles is acute and thus each side adjacent to the right angle less

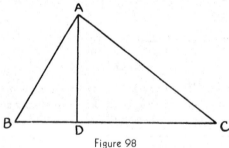

Figure 98

than *q*. The proof, which is almost the same as that for the corresponding theorem in Hyperbolic Geometry, may be supplied by the reader.

Theorem 2. The sum of the angles of any triangle is greater than two right angles.

Let *ABC* (Fig. 98) be any triangle at all. If one of the angles is a right angle, or if two of the angles or all three of them are obtuse, the theorem follows at once. We consider then only the cases in which at least two of the angles, say those at *B* and *C*, are acute. Draw the altitude *AD* from *A* to *BC*. The foot *D* of this altitude will lie on segment *BC*, for, if it did not, the altitude *AD* would have to be, by the lemma of Section 89, both greater than and less than *q* at the same time. It should now be clear, since the angle-sums of right triangles *ADB* and *ADC* exceed two right angles, that so also does the angle-sum of triangle *ABC*.

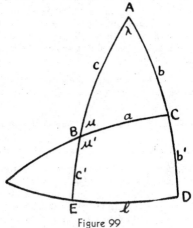

Figure 99

The difference between two right angles and the sum of the angles of a triangle is called the *excess* of the triangle.

Corollary. The sum of the angles of every quadrilateral is greater than four right angles.

In concluding our brief survey of the purely geometric aspects of the Elliptic Plane, we remark that there are, of course, no limiting curves in this geometry, that circles are equidistant curves and that equidistant curves are circles. As a matter of fact, an equidistant curve properly consists of *two* circles, symmetrically situated with regard to the base-line. On the Single Elliptic Plane the two branches of an equidistant curve are connected.[11]

As in Hyperbolic Geometry, there is associated with every right triangle in Elliptic Geometry a Lambert Quadrilateral. Figure 99

[11] See Figure 94.

exhibits the relations between the parts of a right triangle *ABC* and its associated Lambert Quadrilateral *BCDE*. We define c' by the relation $c' + c = q$, μ' is the supplement of μ, and l is the segment of line which subtends an angle λ at its pole. Here again appears, as is easy to show, the series of five associated right triangles.

EXERCISES

1. Prove that the line joining the midpoints of two sides of a triangle is greater than one-half the third side.

2. If two points are each at a distance q from a point O, the line determined by the two points has O for pole. What are the exceptions?

3. Make a list of three or more statements which are true for all three geometries. Make a list of three or more statements which are true for (*a*) Parabolic Geometry only, (*b*) Elliptic Geometry only, (*c*) Hyperbolic Geometry only, (*d*) Parabolic and Elliptic Geometries only, (*e*) Parabolic and Hyperbolic Geometries only, (*f*) Elliptic and Hyperbolic Geometries only.

The following exercises refer to geometry on the Double-Elliptic Plane. They suggest the results to be obtained, with proper modifications, for the Single-Elliptic Plane.

4. Choose the unit of line so that $q = \dfrac{\pi}{2}k$ and the unit of angle so that a right angle measures $\dfrac{\pi}{2}$. Then show that an angle has the measure $\dfrac{x}{k}$ if the length of the segment of line having the vertex for pole and included between the sides has the measure x.

5. Given a triangle, construct the three lines which have its vertices for poles. These lines divide the entire plane into eight triangles. Of these eight, that one is called the *polar* triangle of the given triangle which has its vertices lying in the same relative positions with regard to the corresponding sides of the given triangle as the vertices of the given triangle itself. Prove that, if one triangle is the polar of a second, then the second is the polar of the first. Restrict the discussion in this and the next three examples to triangles which have no angles larger than π.

6. If a triangle *ABC* is lettered in the conventional way, a, b, c designating the measures of the sides and λ, μ, v the measures of the angles, the corresponding parts of the polar triangle being denoted by a', b', c', λ', μ', v', prove that $\lambda + \dfrac{a'}{k} = \pi$, $\mu + \dfrac{b'}{k} = \pi$, etc.

7. Prove that two triangles with the three angles of one equal, respectively, to the three angles of the other are congruent, in other words, that there are no similar triangles in Elliptic Geometry.

8. Show how to construct a triangle, given its three angles.

9. Choose the unit of area so that the area of a digon or biangle with angles $\dfrac{\pi}{2}$ is $k^2\pi$. Prove that the area of a biangle of angle α is $2k^2\alpha$, and that the area of the entire plane is $4\pi k^2$.

10. Prove that the area of a triangle with angles λ, μ, υ is given by $k^2[\lambda + \mu + \upsilon - \pi]$. *Suggestion:* Complete the sides of the triangle, the three lines dividing the plane into eight triangles, congruent in pairs.

91. The Trigonometry of the Elliptic Plane.

We turn briefly to the trigonometry of the Elliptic Plane. The discussion will add a degree of completeness to this hasty survey of Elliptic Geometry and at the same time will serve to introduce an ingenious treatment[12] of Non-Euclidean Trigonometry. The method to be used here is quite different from the one employed in our study of Hyperbolic Trigonometry; the contrast should prove interesting. The reader will find it a profitable exercise to return to the study of the trigonometry of the Hyperbolic Plane and apply to it this alternative treatment.[13]

92. The Trigonometric Functions of an Angle.

For simplicity, we shall define the trigonometric functions only for acute angles. The definitions can be extended later to other angles. This, it will be recognized, is the procedure frequently followed in elementary presentations of Euclidean Trigonometry.

Let θ (Fig. 100), with vertex at O, be any acute angle. From any point P on one side draw PQ perpendicular to the other.

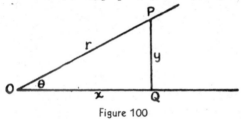

Figure 100

Designate by x, y and r the lengths of segments OQ, QP and OP, respectively. It will be convenient, and will not affect the generality of the results, to choose P so that r is less than q. As will presently appear, the ratios $\dfrac{y}{r}$ and $\dfrac{x}{r}$ are not constant, when the

---

[12] This method is due to P. Mansion. See Mathesis, Second Series, Vol. IV (1894), pp. 180–183.

[13] As a matter of fact, the method was devised first for Hyperbolic Trigonometry by M. Gérard. See footnote, Section 65. It was later modified by Mansion to fit the Elliptic Hypothesis.

position of P varies, as in Euclidean Trigonometry. We propose to show, however, that these ratios approach finite limits as r approaches zero. These limits, which can be shown to be continuous functions of the angle, we define as the *sine* and *cosine* of θ, so that

$$\sin \theta = \lim_{r \to 0} \frac{y}{r}$$

and

$$\cos \theta = \lim_{r \to 0} \frac{x}{r},$$

the other trigonometric functions ensuing in the conventional fashion. We shall need to prove a sequence of theorems.

Theorem 1. As r decreases, the angle OPQ decreases.

Let P_1 and P_2 (Fig. 101) be any two positions of P such that OP_1 is less than OP_2. Draw the perpendiculars P_1Q_1 and P_2Q_2. Then the sum of angles OP_1Q_1 and $Q_1P_1P_2$ is equal to two right angles, while that of angles $Q_1P_1P_2$ and $P_1P_2Q_2$ is greater than two right

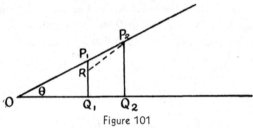

Figure 101

angles. Therefore angle OP_1Q_1 is less than angle OP_2Q_2. It follows as a consequence that the angle at P (Fig. 100) is acute and approaches a right angle as r approaches q.

Theorem 2. As r decreases continuously, so also does y.

As before, let OP_1 (Fig. 101) be less than OP_2. It is clear that P_1Q_1 and P_2Q_2 cannot have the same length, for then angle $P_1P_2Q_2$ would have to be obtuse. Nor can P_1Q_1 be greater than P_2Q_2. For in this case, if Q_1R is measured off on Q_1P_1 equal to Q_2P_2 and P_2 joined to R, it follows that angle RP_2Q_2 is obtuse, contradicting the fact that angle $P_1P_2Q_2$ is acute. We conclude that P_1Q_1 is less than P_2Q_2 and that y decreases as r does.

Theorem 3. As r decreases continuously, so also does the ratio $\dfrac{x}{r}$.

In order to prove this, divide OQ (Fig. 102) into n equal parts, $Q_1, Q_2, Q_3, \ldots, Q_{n-1}$ being the points of division. Draw the perpendiculars to OQ at these points; they will intersect OP in points $P_1, P_2, P_3, \ldots, P_{n-1}$. These divide OP into n parts which are *not* equal, as we shall see. Select any representative set of three consecutive perpendiculars, say P_kQ_k, $P_{k+1}Q_{k+1}$ and $P_{k+2}Q_{k+2}$. Since

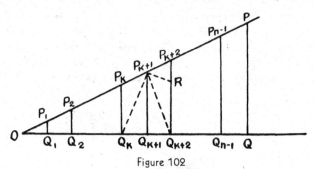

Figure 102

$P_{k+2}Q_{k+2}$ is greater than P_kQ_k, segment $Q_{k+2}R$ can be measured off on $Q_{k+2}P_{k+2}$ equal to Q_kP_k and R joined to P_{k+1}. Then the equality of angles $Q_kP_kP_{k+1}$ and $Q_{k+2}RP_{k+1}$, and of segments P_kP_{k+1} and $P_{k+1}R$, follows from congruence theorems, leading to the conclusion that angle $P_{k+1}P_{k+2}R$ is greater than angle $P_{k+1}RP_{k+2}$. Consequently, since triangle $P_{k+2}P_{k+1}R$ is of limited size, side $P_{k+1}R$ is greater than side $P_{k+1}P_{k+2}$, or, what is the same thing, P_kP_{k+1} is greater than $P_{k+1}P_{k+2}$.

Turning our attention now to the ratio $\dfrac{x}{r}$, we start with r equal to OP_1. In this case the ratio is less than unity. As r becomes successively OP_2, OP_3, $\ldots$, OP, the ratio $\dfrac{x}{r}$ increases, since x receives equal increments and r decreasing ones. Then for any two positions of P, for which the x's are commensurable, the corresponding ratios are unequal, and that is the greater for which r is greater. The same conclusion can be reached if the x's are incommensurable by the use of limiting processes. Thus we conclude that the ratio $\dfrac{x}{r}$ increases continuously as r increases, and hence decreases as r does.

For later use, we call attention to the fact that if, assuming that OP and OQ (Fig. 102) are of length q, the angles at P and Q being right angles, we start with the Lambert Quadrilateral $PQQ_{n-1}P_{n-1}$, we can easily show that the ratio $\dfrac{PP_{n-1}}{QQ_{n-1}}$ increases as QQ_{n-1} does and decreases as QQ_{n-1} decreases.

Theorem 4. As r decreases continuously, the ratio $\dfrac{y}{r}$ increases.

This time, divide OP (Fig. 103) into n equal segments, with points of division P_1, P_2, P_3, ... , P_{n-1}, and draw the perpendiculars P_1Q_1, P_2Q_2, P_3Q_3, ... , $P_{n-1}Q_{n-1}$ to OQ. Select, as before, any representative set of three consecutive perpendiculars, P_kQ_k, $P_{k+1}Q_{k+1}$

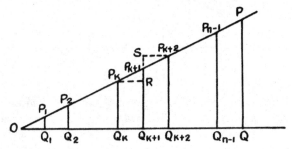

Figure 103

and $P_{k+2}Q_{k+2}$. From P_k draw P_kR perpendicular to $P_{k+1}Q_{k+1}$. On $Q_{k+1}P_{k+1}$ produced through P_{k+1} measure off $P_{k+1}S$ equal to $P_{k+1}R$ and join P_{k+2} to S. Since triangles P_kRP_{k+1} and $P_{k+1}SP_{k+2}$ are congruent, angle $P_{k+1}SP_{k+2}$ is a right angle. Designate the common length of $P_{k+1}S$ and $P_{k+1}R$ by d. Then

$$P_{k+2}Q_{k+2} < SQ_{k+1} = P_{k+1}Q_{k+1} + d$$

and

$$P_kQ_k < RQ_{k+1} = P_{k+1}Q_{k+1} - d.$$

Therefore

$$P_{k+2}Q_{k+2} - P_{k+1}Q_{k+1} < d$$

and

$$P_{k+1}Q_{k+1} - P_kQ_k > d,$$

so that y receives decreasing increments as r receives equal ones. Arguing as before, we are led to the conclusion that the ratio $\frac{y}{r}$ decreases continuously as r increases, and increases as r decreases.

We are now ready to show that the ratios under consideration actually approach limits as r approaches zero. Theorem 3 convinces us that, as r decreases, the ratio $\frac{x}{r}$, since it is always positive and decreasing, approaches a limit which may be a positive proper fraction or zero. On the other hand, Theorem 4 alone fails to give us sufficient information to conclude that $\frac{y}{r}$ also approaches a limit. That this is, however, the case can be shown as follows:

Starting with the angle used in the other drawings, and employing the same letters, construct (Fig. 104) OT perpendicular to OQ at O, and draw PR perpendicular to OT. Designate by x' and y' the lengths of segments PR and OR, respectively. Then, since angle POR is

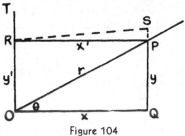

Figure 104

acute, the theorems just proved can be applied to it, as well as to angle POQ. Figure $OQPR$ is a Lambert Quadrilateral. Consequently x is greater than x' and y is less than y'. Since, as r decreases, $\frac{x'}{r}$ is positive and increasing, and $\frac{x}{r}$ is always greater than $\frac{x'}{r}$, it is now clear that the limit of $\frac{x}{r}$ is not zero. It immediately follows that $\frac{y'}{r}$ also approaches a positive limit less than unity from above, and, since $\frac{y}{r}$ is always less than $\frac{y'}{r}$ and continuously increases as r decreases, it must approach a finite limit, which is less than unity, from below.

93. Properties of a Variable Lambert Quadrilateral.

Figure 104, of the last section, is to play an important part in the investigations to follow. Of particular interest to us are the consequences of fixing side y' of the Lambert Quadrilateral $OQPR$ and allowing x to vary. Then x', y, r, θ and the obtuse angle RPQ will be variable also. We wish to consider, in particular, the character of the variation of the ratio $\dfrac{x'}{x}$ as x approaches zero, y' being fixed. The following theorem serves to introduce an important function.

Theorem. If the sides of a Lambert Quadrilateral are x, y, x', y', where x and x', y and y' are pairs of opposite sides, x' and y including the obtuse angle, and if y' is kept fixed while x is allowed to approach zero, then the ratio $\dfrac{x'}{x}$ decreases and approaches a finite limit $\varphi(y')$.

In order to prove this theorem, we shall need two lemmas.

Lemma 1. If two Lambert Quadrilaterals $ABCD$ and $A'B'C'D'$ have their obtuse angles at A and A', and if AB and $A'B'$ are equal, while BC is less than $B'C'$ and each is less than q, then angle BAD is less than angle $B'A'D'$.

The *reductio ad absurdum* proof is left to the reader.

Lemma 2. If $ABCD$ is a Lambert Quadrilateral, obtuse-angled at A, then, if AB is kept fixed and BC allowed to decrease continuously and approach zero, the ratio $\dfrac{AD}{BC}$ increases.

Divide BC (Fig. 105) into n equal parts, C_1, C_2, ... , C_{n-1} being the points of division, draw the perpendiculars to BC at these points, and to these lines draw the perpendiculars AD_1, AD_2, ... , AD_{n-1} from A to form Lambert Quadrilaterals. From Lemma 1, it is clear that the obtuse angle decreases as BC does. Let ABC_kD_k, $ABC_{k+1}D_{k+1}$, $ABC_{k+2}D_{k+2}$ be three representative successive quadrilaterals. Produce AD_k to cut $C_{k+1}D_{k+1}$ in E, C_kD_k to cut AD_{k+1} in F,

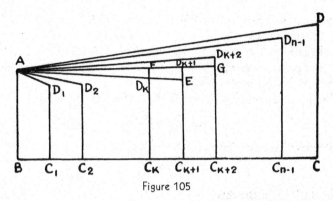

Figure 105

and AD_{k+1} to cut $C_{k+2}D_{k+2}$ in G. Since quadrilaterals $D_{k+1}C_{k+1}C_kF$ and $D_{k+1}C_{k+1}C_{k+2}G$ are congruent, FD_{k+1} is equal to $D_{k+1}G$. Furthermore, AG is greater than AD_{k+2} and AD_k less than AF. Then

$$AD_{k+1} + FD_{k+1} > AD_{k+2},$$

so that

$$AD_{k+1} - AF > AD_{k+2} - AD_{k+1},$$

and finally

$$AD_{k+1} - AD_k > AD_{k+2} - AD_{k+1}.$$

Thus, starting with quadrilateral ABC_1D_1, as BC_1 takes on equal increments, AD_1 takes on decreasing ones. Arguing as in Section 92, we conclude that the ratio $\dfrac{AD}{BC}$ decreases as BC increases and hence increases as BC decreases.

Returning to the theorem, we point out first that, as a consequence of the remark made at the end of the proof of Theorem 3 of the last section, the ratio $\dfrac{x'}{x}$ decreases as x approaches zero. It therefore approaches a limit less than unity. We shall show that it is not zero.

In Figure 104, draw RS perpendicular to QP produced. As a consequence of Lemma 2, the ratio $\dfrac{RS}{x}$ increases as x decreases. But, since RP is greater than RS, it follows that the ratio $\dfrac{x'}{x}$ is always greater than $\dfrac{RS}{x}$. We deduce that the ratio $\dfrac{x'}{x}$ has a limit which is greater than zero. This limit depends upon the length of y' and we designate it by $\varphi(y')$. It should be noted that, at least as far as this discussion is concerned, y' is restricted to lie between zero and q. As y' approaches zero, $\varphi(y')$ approaches unity; as y' approaches q, $\varphi(y')$ approaches zero.

We have shown that, for quadrilateral $OQSR$ (Fig. 104),

$$\varphi(y') > \frac{RS}{x}.$$

Applying this result to quadrilateral $OQPR$, we have

$$\varphi(y) > \frac{x'}{x}$$

and thus obtain the relation

$$\varphi(y) > \frac{x'}{x} > \varphi(y'),$$

for which we shall have use later. Since y' is greater than y, it shows that as x increases $\varphi(x)$ decreases.

94. The Continuity of the Function $\varphi(x)$.

Let $ABCD$ (Fig. 106) be a Lambert Quadrilateral with obtuse angle at C. Designate the lengths of segments AB, BC and DA by x, u and v, respectively. We have shown that, if x is fixed and v allowed to approach zero, the ratio $\frac{u}{v}$ approaches from above a limit less than unity, which depends upon the length of x and is

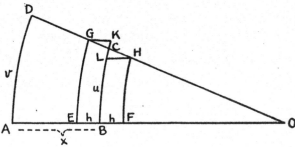

Figure 106

denoted by $\varphi(x)$. We wish to prove that $\varphi(x)$ is a continuous function.

In order to do this, produce DC through C until it cuts AB produced in O. The point O is a pole of line AD. Measure off on AO, in both directions from B, segments BE and BF of length h, and at E and F draw perpendiculars to AO, cutting DO in G and H. From G and H draw GK and HL perpendicular to BC. From Theorem 3 of Section 92, we have

$$\frac{CK}{CG} < \frac{CB}{CO},$$

which can be written

$$\frac{BK - u}{v} < \frac{u}{v} \cdot \frac{CG}{CO},$$

or

$$\frac{GE}{v} \cdot \frac{BK}{GE} - \frac{u}{v} < \frac{CG}{CO} \cdot \frac{u}{v}.$$

Then, as v is allowed to approach zero, the line DO rotating about O, we have, proceeding to the limit,

$$\frac{\varphi(x - h)}{\varphi(h)} - \varphi(x) \leqq \frac{h}{q - x} \cdot \varphi(x),$$

or

$$\varphi(x - h) - \varphi(x)\varphi(h) \leqq \frac{h}{q - x}\varphi(x)\varphi(h). \tag{1}$$

Again, starting with the relation

$$\frac{CL}{CH} < \frac{CB}{CO},$$

we obtain, in exactly the same way, the relation

$$\varphi(x)\varphi(h) - \varphi(x + h) \leqq \frac{h}{q - x}\varphi(x)\varphi(h). \tag{2}$$

Combining (1) and (2), we find that

$$\varphi(x - h) - \varphi(x + h) \leqq \frac{2h}{q - x}\varphi(x)\varphi(h) < \frac{2h}{q - x}.$$

Thus, no matter what value x has in the interval under consideration, a value of h can be found such that the difference of $\varphi(x - h)$ and $\varphi(x + h)$ will be less than any assigned positive quantity, however small, and the continuity of $\varphi(x)$ is assured.

95. An Important Functional Equation.

We turn next to a figure very much like the preceding one, and use it to derive a fundamental condition satisfied by the function $\varphi(x)$.

Let $ABCD$ (Fig. 107) be a Lambert Quadrilateral, obtuse-angled at C. Designate the lengths of segments AB, BC and DA by x, u and v, respectively, and produce AB and DC to meet at O.

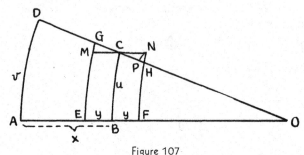

Figure 107

Measure off on AO segments BE and BF equal to y and draw the perpendiculars to AO at E and F. Let them cut DO at G and H. This time draw the perpendicular to CB at C and let it cut EG in M and FH produced in N. It is easy to see that CM and CN are equal. Finally, since CH is greater than GC, segment CP equal to GC can be measured off on CH and P joined to N. Then NP is equal to GM. As v is allowed to approach zero, the angles GMC and HNC approach right angles, angle PNH becomes infinitesimal and so do segments NH and PH. It can be shown that PH is an infinitesimal of higher order than v, and we shall assume this here. Then since

$$(NF - HF) - (GE - ME) = NH - GM = NH - NP < PH,$$

it follows that

$$\lim_{v \to 0} \left[\frac{NF}{u} \cdot \frac{u}{v} - \frac{HF}{v} - \frac{GE}{v} + \frac{ME}{u} \cdot \frac{u}{v} \right] = 0,$$

or

$$\varphi(y)\varphi(x) - \varphi(x+y) - \varphi(x-y) + \varphi(y)\varphi(x) = 0,$$

or finally

$$\varphi(x+y) + \varphi(x-y) = 2\varphi(x)\varphi(y).$$

96. The Function $\varphi(x)$.

The discovery that the function $\varphi(x)$ satisfies the condition arrived at in the preceding section together with the facts that it is less than or equal to unity, decreases as x increases, is equal to unity when x is zero and to zero when x is equal to q, suggests a resemblance to the familiar function $\cos x$. We propose to show that[14]

$$\varphi(x) = \cos \frac{x}{k},$$

where k is a constant depending upon the unit of length.

Choose, to begin with, a particular value of x, say x', the choice to be governed only by restrictions already imposed on x, as they apply to the discussion which follows. Then $\varphi(x)$ has a value between zero and unity and is thus equal to the cosine of some argument. Whatever this argument is, it can be represented by $\frac{x'}{k}$, if k is properly chosen. Consequently

$$\varphi(x') = \cos \frac{x'}{k}.$$

Since

$$\varphi(px) = 2\varphi[(p-1)x]\,\varphi(x) - \varphi[(p-2)x]$$

and

$$\cos (px) = 2 \cos [(p-1)x] \cos x - \cos [(p-2)x],$$

regardless of the value of p, it easily follows by mathematical induction that

$$\varphi(nx') = \cos \frac{nx'}{k},$$

where n is any positive integer. Then, in a similar way, we find that

$$\varphi\left(\frac{nx'}{2^m}\right) = \cos \frac{nx'}{2^m k},$$

where m is any positive integer. Thus the relation

$$\varphi(x) = \cos \frac{x}{k}$$

holds for every value of x, within the interval under consideration, of the form $\frac{nx'}{2^m}$. That it holds for every other value of x in the interval follows from the continuity of the functions $\varphi(x)$ and $\cos x$ and from the fact that $\frac{nx'}{2^m}$ can, by proper choice of m and n, be made to differ from any such value of x by as small an amount as we please.

[14] The reference here is, of course, to the generalized definition of $\cos x$. See Appendix II.

97. The Relations among the Parts of a Right Triangle.

We have now reached the point where we are ready to derive the basic relations of Elliptic Trigonometry, those which relate the measures of the sides and angles of a right triangle.

Let ABC (Fig. 108) be any right triangle with the angle at C the right angle, and designate the measures of the sides in the customary way by a, b and c. Again we shall restrict the size of the figure so that the segments involved will be less than q. Extension of the results to unrestricted figures leads to no great difficulty.

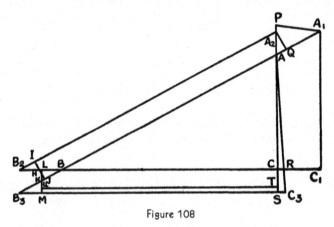

Figure 108

On BA, produced through A, mark off a segment AA_1 and draw A_1C_1 perpendicular to BC, produced through C. On CA and CB produced measure off CA_2 and CB_2 equal to C_1A_1 and C_1B, respectively, and join A_2 and B_2. Triangles A_1BC_1 and A_2B_2C are congruent. Next produce AB through B, measure AB_3 equal to A_1B, and construct at B_3 angle AB_3C_3 equal to angle A_1BC. If B_3C_3 is made equal to BC_1 and A joined to C_3, triangles A_1BC_1 and AB_3C_3 will be congruent. It is easy to see that BB_2 is equal to CC_1 and BB_3 to AA_1.

Through the midpoint H of B_2B draw HI perpendicular to A_2B_2. If BJ is measured off on BB_3 equal to B_2I and H joined to J, triangles HB_2I and HBJ will be congruent, points I, H and J will be collinear, and HJ will be perpendicular to B_3A. In a similar way LKM, through the midpoint K of BB_3 and perpendicular to B_2C and B_3C_3, can be drawn.

Since angle BA_1C_1 is greater than angle BAC, AC_3 will intersect BC_1 between C and C_1. Produce AC to cut B_3C_3 in S. Then AS is greater than AC_3, which in turn is equal to A_2C. Thus CS is greater than A_2A. Consequently segment CT, equal to AA_2, can be marked off on CS, the point T lying between C and S. Draw the perpendicular TU from T to LM. Finally, to complete this elaborate figure, draw A_1P perpendicular to AC and A_2Q perpendicular to AB.

If we allow AA_1 to approach zero, we see that

$$\lim \frac{A_1P}{AA_1} = \lim \frac{A_2Q}{AA_2}.$$

Similarly

$$\lim \frac{LK}{BK} = \lim \frac{JH}{BH},$$

so that

$$\lim \frac{LM}{BB_3} = \lim \frac{IJ}{BB_2}.$$

Then, from the first and third of these equations, we obtain by division

$$\lim \frac{A_1P}{LM} = \lim \frac{A_2Q}{IJ} \lim \frac{CC_1}{AA_2},$$

or, what is the same thing,

$$\lim \frac{A_2Q}{IJ} = \lim \frac{A_1P}{CC_1} \lim \frac{AA_2}{LM}.$$

We shall consider, in turn, each of the three limits involved.

First, if we apply the last relation obtained in Section 93 to the Lambert Quadrilateral A_2IJQ, we find that

$$\varphi(JQ) < \frac{A_2Q}{IJ} < \varphi(IA_2).$$

Proceeding to the limit, JQ and IA_2 both approach AB and thus, on account of the continuity of the function $\varphi(x)$, both $\varphi(JQ)$ and $\varphi(IA_2)$ approach $\varphi(AB)$. Hence

$$\lim \frac{A_2Q}{IJ} = \varphi(AB),$$

and, in exactly the same way by use of quadrilateral A_1PCC_1,

$$\lim \frac{A_1P}{CC_1} = \varphi(AC).$$

The third limit is not quite so easily disposed of. However, since

A_2C is equal to AC_3 and AC is less than AR, we observe that RC_3 is less than AA_2, and consequently

$$\frac{AA_2}{LM} > \frac{RC_3}{LM} > \varphi(MC_3).$$

Also

$$\frac{AA_2}{LM} < \frac{AA_2}{LU} = \frac{CT}{LU} < \varphi(UT).$$

Then, since MC_3 and UT both approach BC, as AA_1 approaches zero,

$$\lim \frac{AA_2}{LM} = \varphi(BC).$$

Thus we conclude that

$$\varphi(AB) = \varphi(AC)\varphi(BC),$$

or

$$\cos \frac{c}{k} = \cos \frac{a}{k} \cdot \cos \frac{b}{k}. \tag{1}$$

This is the relation connecting the three sides of a right triangle. It leads immediately to the important conclusion that the Pythagorean Formula holds for infinitesimal right triangles in Elliptic Geometry. From this fact, it follows that the trigonometric functions of angles, as defined in Section 92, are connected by the familiar relationships of Euclidean Trigonometry, namely,

$$\sin^2 \theta + \cos^2 \theta = 1,$$
$$\sec^2 \theta - \tan^2 \theta = 1, \text{ etc.}$$

To obtain a second formula relating parts of a right triangle ABC, produce CA (Fig. 109) through A any convenient distance u to point P, and draw the perpendicular PQ from P to BA produced. Designate the measure of PQ by v and of AQ by w. Finally draw BP. Then

$$\cos \frac{BP}{k} = \cos \frac{a}{k} \cos \frac{b+u}{k}$$

$$= \frac{\cos \dfrac{c}{k}}{\cos \dfrac{b}{k}} \left[\cos \frac{b}{k} \cos \frac{u}{k} - \sin \frac{b}{k} \sin \frac{u}{k} \right]$$

and

$$\cos \frac{BP}{k} = \cos \frac{v}{k} \cos \frac{c+w}{k}$$

$$= \frac{\cos \dfrac{u}{k}}{\cos \dfrac{w}{k}} \left[\cos \frac{c}{k} \cos \frac{w}{k} - \sin \frac{c}{k} \sin \frac{w}{k} \right],$$

so that

$$\frac{\tan \dfrac{b}{k}}{\tan \dfrac{c}{k}} = \frac{\tan \dfrac{w}{k}}{\tan \dfrac{u}{k}}.$$

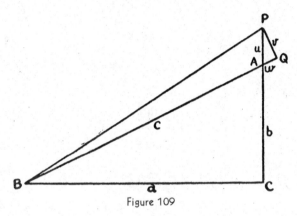

Figure 109

If u is allowed to approach zero, the right-hand member of this equation approaches $\cos A$, so that we have

$$\cos A = \frac{\tan \dfrac{b}{k}}{\tan \dfrac{c}{k}}, \tag{2}$$

the formula relating an angle — not the right angle C — to its including sides.

It is no great task to convert this into a third formula, for we obtain at once

$$\sin^2 A = 1 - \frac{\tan^2 \frac{b}{k}}{\tan^2 \frac{c}{k}}$$

$$= \frac{\sec_2 \frac{c}{k} - \sec^2 \frac{b}{k}}{\tan^2 \frac{c}{k}}$$

$$= \frac{1 - \cos^2 \frac{a}{k}}{\sin^2 \frac{c}{k}}$$

or

$$\sin A = \frac{\sin \frac{a}{k}}{\sin \frac{c}{k}}. \tag{3}$$

The other three formulas,

$$\cot A = \frac{\sin \frac{b}{k}}{\tan \frac{a}{k}}, \tag{4}$$

$$\cos A = \cos \frac{a}{k} \sin B, \tag{5}$$

$$\cos \frac{c}{k} = \cot A \cot B, \tag{6}$$

are then easily obtained.

The reader will recognize these formulas as those for the spherical right triangle for a sphere of radius k.

VIII

THE CONSISTENCY OF THE NON-EUCLIDEAN GEOMETRIES

"In respect of soundness — inner consistency — self-compatibility — logical concordance among the parts of each — the three geometries are on exactly the same level, and the level is the highest that man has attained. The three doctrines are equally legitimate children of one spirit, — the geometrizing spirit, which Plato thought divine, — and they are immortal. Work inspired and approved by the muse of intellectual harmony cannot perish — it is everlasting." — CASSIUS J. KEYSER

98. Introduction.

Sooner or later, in the study of Non-Euclidean Geometry, the following question is certain to arise: which of the three geometries is the "true" geometry, or, in other words, which geometry actually describes our physical space? In this connection it is hoped that enough has already been said about the Kantian space philosophy to convince the reader of its weakness and, from the standpoint of the geometer at least, to discredit it entirely. That space is an idea existing *a priori* in the minds of humans and without which there would be no space phenomena as we know them, is a viewpoint no longer considered satisfactory. Geometry, when applied to space, becomes experimental in character and should be looked upon, as Gauss remarked, from the same standpoint as mechanics. The space which we recognize through the organs of sense, composed as it is of a multitude of discrete impressions, lacks much of being a mathematical continuum.

Thus we conclude that there is no point to the question at the beginning of this section. One geometry is no more " true" than the others. As a matter of fact, when a geometry is applied to space, a postulate of parallels becomes an empirical law, like the law of falling bodies, which at best appears to describe very well things as they seem to be. As pointed out in the first chapter, the application of geometry to physical space is but an attempt to supply a body of logical doctrine which will correlate, with sufficient accuracy for all practical purposes, the data of observation and experiment. And it is true that any one of the three geometries does this about as well as either of the others. If indirect measurements were made and engineering projects constructed on the basis of Hyperbolic or Elliptic Geometry, the results would be as satisfactory as those obtained under the Euclidean Hypothesis. Indeed, the differences would very likely not be perceptible within this modicum of space to which we are physically restricted. But even if there were accruing evidence that one of the Non-Euclidean Geometries describes our space in some respects more precisely than Euclidean, the latter would still continue to be largely used because of its comparative simplicity.

As pointed out in Section 74, we can never ascertain that space is Euclidean, even if that is the case, because of experimental errors which cannot be entirely eliminated. But it is not altogether impossible that improved instruments and new methods, leading, for example, to the fixing of an upper limit to the space constant, may eventually enable us to assert definitely that the universe is essentially Non-Euclidean. However, to such an end there are very effective barriers. One of these is the fact that the methods of indirect measurement, as well as the very instruments of measurement themselves, must be fashioned upon one or other of the geometries as basis. Conclusions of the kind sought, based upon results obtained by the use of such methods and instruments, would hardly be convincing.

But there is another question which must be disposed of in another way: are the Non-Euclidean Geometries consistent? As far as we ourselves have gone with the development of the two geometries, no contradiction has been discovered. As a matter of fact, no contradiction has ever been discovered by anyone. However, are we sure, as investigations are continued, that contradictions will not arise to show that the postulates of one or the other are incompatible?

Many proofs of the fact that the Non-Euclidean Geometries are consistent have been given, from the earliest one, presented by Beltrami in 1868 and already mentioned,[1] to the significant and far-reaching, though much more elaborate, demonstrations of Cayley and Klein. Some proofs, like Beltrami's, depend upon the representations of the Non-Euclidean Geometries upon Euclidean surfaces of constant curvature; others are analytical in character and obtain their objective by showing that the analytical representations lead to sets of consistent equations. Still others appeal to the methods of projective geometry. The proof[2] which we offer here is synthetic, and depends only upon simple geometric concepts. No knowledge of projective geometry or of the geometry of surfaces is required.

We shall prove that the postulates of Hyperbolic Plane Geometry form the basis of a consistent logical system by directing attention to the set of circles in Euclidean Geometry which cut a fixed circle at right angles. We shall show that the geometry of this family of circles, when properly interpreted, presents a complete analogue of geometry on the Hyperbolic Plane, definition by definition, postulate by postulate, proposition by proposition. As a consequence, we shall be able to infer that Hyperbolic Geometry is consistent. For, if any contradiction were ever encountered, there would be a corresponding inconsistency at the corresponding point in the geometry of the family of circles. But the latter is primarily Euclidean Geometry. Hence it would have to follow that Euclidean Geometry itself is not consistent. The plan can readily be extended to solid geometry.

It will be recognized that this test for consistency is comparative.

---

[1] See Section 33.

[2] The method is due to Poincaré. See the English translation of his *Science and Hypothesis* by W. J. Greenstreet (London, 1905). We follow the more detailed treatment given by H. S. Carslaw, first in the *Proceedings of the Edinburgh Mathematical Society*, Vol. XXVIII (1910), pp. 95–120, then in the appendix to his translation of Bonola: *Non-Euclidean Geometry* (Chicago, 1912), and finally in his *Elements of Non-Euclidean Plane Geometry and Trigonometry* (London, 1916). It is an extension and elaboration of the treatment of the subject by Wellstein in Weber and Wellstein's *Encyclopädie der Elementar-Mathematik*, Vol. ii, pp. 28–83 (Leipzig, 1907).

But so are all of the others. In fact, no absolute method for testing the consistency of a set of assumptions has ever been found.[3]

99. The Geometry of the Circles Orthogonal to a Fixed Circle.

Let us consider a geometry in which the circles orthogonal[4] to a fixed circle, called the *fundamental circle*, play the rôles of lines. This geometry is to be basically Euclidean, but we shall interpret it in such a way that it will satisfy the postulates of Hyperbolic Geometry.

Any circle may be chosen as the fundamental circle. Every point inside this circle is regarded as a point of our geometry; the points on the circumference and outside the circle are not considered as points of the geometry at all. Those arcs of circles cutting the fundamental circle orthogonally which lie within this circle consti- tute the lines of the geometry. Since they are not lines in the strict sense, we shall call them *nominal lines*.[5] It will be recalled that one and only one circle can be constructed through two points within a circle and orthogonal to it, provided, of course, we count straight lines through the center of a given circle as belonging to the aggre- gate of circles orthogonal to it. Thus it is clear that nominal lines satisfy a prime requisite of lines in Hyperbolic Geometry, namely, that a line is determined by two of its points. Furthermore, the axioms of order hold. Since a nominal line is not closed, it is always possible to designate one of three points on it as lying *between* the other two. On account of the fact that two circles which intersect, and are both orthogonal to a third circle, have one point of inter- section inside and the other outside the third circle, it follows that if two nominal lines intersect, they intersect in one point only.

We shall define the angle of intersection of two intersecting nominal lines as the angle of intersection of the tangent lines, at the point of intersection, to the circles with which the lines coincide. Thus two intersecting lines form four angles, equal in pairs and supplementary in pairs. Two nominal lines are said to be perpendic-

[3] A test for the consistency of a geometry does not depend necessarily upon com- parison with another *geometrical* system. The ideas of geometry can, for example, be transferred to the realm of numbers. See, in this connection, Hilbert's proof of the consistency of the axioms of Euclidean Geometry (*Grundlagen der Geometrie*, 7th ed., p. 34, or Townsend's translation, p. 27) and the proof offered by Veblen and Young for the miniature mathematical system described in their *Projective Geometry*, Vol. i, p. 3 (Boston, 1910). In the latter place there will also be found references to the literature on the subject of the possibility of an absolute test for consistency.

[4] See the Appendix for a brief account of the theory of orthogonal circles.

[5] As suggested by Carslaw.

ular when they intersect at right angles. One and only one circle can be drawn through a point and orthogonal to two intersecting circles. Hence, through a given point, there can be drawn in our geometry one and only one nominal line perpendicular to a given nominal line.

But two nominal lines may not intersect at all. We distinguish between two cases. If the circles with which two nominal lines coincide do not intersect one another, we call the lines *non-intersecting*. If the circles are tangent to one another at a point on the fundamental circle, we call the nominal lines *parallel*. It is easy to see that two nominal lines have a common perpendicular if and only if they are non-intersecting. Through a given point two nominal lines can be constructed parallel to a given nominal line. These parallels separate all of the nominal lines through the given point into two sets: those which intersect the given nominal line and those which do not. Parallel nominal lines may be regarded as intersecting at a zero angle.

The sum of the angles of a nominal triangle, a triangular figure the sides of which are nominal lines, is less than two right angles. To prove this, let *ABC* (Fig. 110) be such a triangle, the funda-

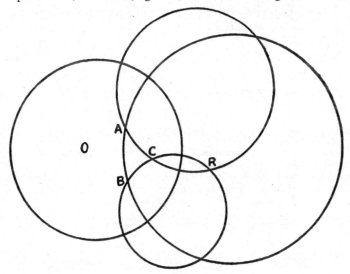

Figure 110

mental circle having its center at O. Complete the circles with which the nominal lines forming the sides of the triangle coincide. In particular let the circles AC and BC intersect again at R. Invert[6] the entire figure with R as center of inversion. Circles AC and BC will invert into intersecting straight lines $A'C'$ and $B'C'$ (Fig. 111). The fundamental circle will invert into a circle orthogonal to lines

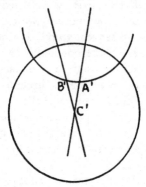

Figure 111

$A'C'$ and $B'C'$ and hence with center at C'. Circle AB will invert into a circle orthogonal to the inverse of the fundamental circle and thus with center outside it. Obviously the angle sum of the triangular figure $A'B'C'$ is less than two right angles, and, since angles are preserved under inversion, the same is true of triangle ABC. The reader will have no difficulty in constructing a figure which may be regarded as a triangle with each angle a zero angle; each side will be parallel to the other two in opposite directions.

EXERCISES

1. Prove that the two nominal lines through a given nominal point, parallel to a given nominal line, make equal angles with the nominal line through the given point and perpendicular to the given line. We thus have in the geometry of the nominal lines an analogue for the angle of parallelism in Hyperbolic Geometry.

2. Prove that such an angle of parallelism is always acute.

100. The Nominal Length of a Segment of Nominal Line.

Our objective is, as has been explained, to prove that the entities of the geometry which we are describing, when assigned the roles

---

[6] See the Appendix for a discussion of the elements of the theory of inversion.

of entities in Hyperbolic Geometry, satisfy the postulates which form the foundation of that geometry. Thus far the evidence has accumulated rapidly. In fact, we have reached a point where we can announce that every proposition of Hyperbolic Geometry which involves *only properties of angles* has its analogue in the geometry of the nominal lines.

But we have not yet given any attention to the idea of the *length* of a segment of nominal line. Until the *nominal length* of a *nominal segment* has been defined for our geometry, it will contain no counterparts for any of the beautiful metric theorems which so vividly characterize Hyperbolic Geometry.

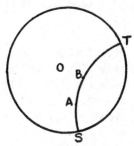

Figure 112

There are three requisites for nominal lines which must be kept in mind in selecting a definition for the nominal length of a nominal segment, if the analogy to Hyperbolic Geometry is to be preserved:

1. A nominal line must be infinitely long.

2. If A, B and C are any three points at all on a nominal line, then nominal length AB + nominal length BC = nominal length AC, sense being taken into account.

3. The nominal length of a nominal segment must be unchanged by displacement.

The following definition satisfies these requirements. Let the circle with which the nominal line AB coincides cut the fundamental circle in points S and T (Fig. 112). Then we define the *nominal length* of the *nominal segment* AB as

$$\log_e \frac{AT}{AS} \Big/ \frac{BT}{BS},$$

where AT, AS, BT, BS designate chords of the circle.[7]   A segment AB will be a unit segment when this logarithm is unity, i.e., when

$$\frac{AT}{AS} \cdot \frac{BS}{BT} = e.$$

The nominal length of the nominal segment AB may be defined more generally as

$$k \log_e \frac{AT}{AS} \Big/ \frac{BT}{BS},$$

where k is a parameter, the choice of which determines the unit of length. Or, what is the same thing, the base for the logarithms may be taken as a instead of e, a change in a amounting to a change in the unit of length. For simplicity, we shall in what follows, unless otherwise indicated, take k equal to unity and the base for the logarithms equal to e.

The reader will not find it hard to show, as a consequence of this definition, that the nominal length of AB becomes infinite as A approaches S or B approaches T. Nor will he have any difficulty in proving that the second condition is satisfied. But the third is another matter, for we have yet to tell what we mean by a displacement in the geometry we are devising.

For the purpose of comparing figures, Euclid made use of superposition, assuming that figures can be displaced or moved in a plane without any change in size or shape. We have used this principle, or what is equivalent to it, in our studies of Hyperbolic Geometry and Elliptic Geometry. We choose to treat congruence and allied topics in the geometry under investigation by this familiar method. Let us consider how such displacements can effectively be brought about.

101. Displacement by Reflection.

In any one of the three geometries, any plane figure can be *reflected* with regard to any line in its plane, called the *axis of reflection*. If P is any point of the figure, and the perpendicular PQ is drawn to the axis and produced to P' so that QP' is equal to PQ, then P is said to have been *reflected* and P' is its *image*. This simple transformation

---

[7] As evidence that there is nothing forced about the representation of length as a logarithm, the reader is referred to Exercise 1, Section 66. The relation of this definition to the cross-ratio of four points will be recognized by the advanced student.

does not change the size and shape of a figure; lengths and angles are preserved.

We are interested in reflection because it is a transformation which can be used to move a plane figure from one position in the plane to any other. To prove this it will suffice to show that a line segment AB (Fig. 113) can, by reflections with regard to properly chosen axes, be made to coincide with any equal segment $A'B'$.

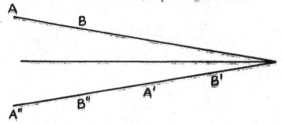

Figure 113

If segments AB and $A'B'$ do not lie in the same straight line and are neither parallel nor segments of non-intersecting lines, let AB be reflected about the bisector of the angle formed by producing lines AB and $A'B'$ until they meet. Its image $A''B''$ will be collinear with $A'B'$ and have the same length. The modification in the method of determining the axis of reflection to obtain the same effect when AB and $A'B'$ are parallel segments or segments of non-intersecting lines, will suggest itself to the reader. If, at this point, $A'B'$ and $A''B''$ have the same sense and do not coincide, segment $A''B''$ must be reflected about its perpendicular bisector so that its sense will be opposite that of $A'B'$. Then a final reflection about the common perpendicular bisector of $A'A''$ and $B'B''$ will bring the segments into coincidence.

The identifying characteristics of reflection are: (1) The segment joining a point to its image is perpendicular to the axis of reflection and bisected by it. (2) A segment of line and its image have equal lengths and, if produced, meet on the axis of reflection, provided they intersect, and make equal angles with it.

102. Displacement in the Geometry of the Nominal Lines.

We are ready now to disclose the analogue of reflection in the geometry which we are constructing. Once it has been described, we shall be able to study the problem of displacement for that geometry. We shall show that inversion of any figure in our geometry, with regard to the circle with which any nominal line coincides as circle of inversion, is a transformation possessing all of the characteristics of reflection about that line, our definitions of nominal length and angle being taken into account.

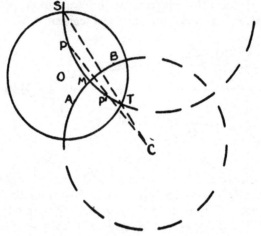

Figure 114

Let the circle with center O (Fig. 114) be the fundamental circle and AB any nominal line. Then AB coincides with an arc of a circle which is orthogonal to the fundamental circle; let the center of that circle be C. If P is any nominal point and P' its inverse with regard to the circle with center C, then the nominal line PP' coincides with a circle which is orthogonal not only to the fundamental circle but also to the circle with center C. In other words, the nominal line PP' is perpendicular to the nominal line AB. We have next to show that the nominal segment PP' is bisected by nominal line AB.

Let the circle on which the nominal line PP' lies cut the fundamental circle in points S and T and the nominal line AB in point M. It is easy to see that S and T are inverse points with regard to the circle with center C. Then

$$\frac{PT}{P'S} = \frac{CT}{CP'}$$

and

$$\frac{P'T}{PS} = \frac{CP'}{CS},$$

so that

$$\frac{PT}{P'S} \cdot \frac{P'T}{PS} = \frac{CT}{CS}.$$

But, since

$$\frac{MT}{MS} = \frac{CM}{CS} = \frac{CT}{CM},$$

we observe that

$$\frac{MT^2}{MS^2} = \frac{CT}{CS},$$

and thus that

$$\frac{PT}{P'S} \cdot \frac{P'T}{PS} = \frac{MT^2}{MS^2}.$$

Therefore

$$\log \frac{PT}{PS} \Big/ \frac{MT}{MS} = \log \frac{MT}{MS} \Big/ \frac{P'T}{P'S}$$

and the nominal lengths of nominal segments PM and MP' are equal.

We conclude that the inversion of a nominal point, with regard to the circle upon which a nominal line lies, may be regarded as a reflection of the point with the nominal line used as axis of re-

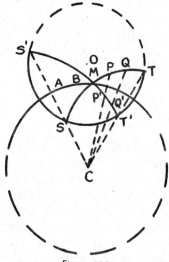

Figure 115

flection. We shall determine next how the inverse of a segment of nominal line, with regard to such a circle, is related to that segment.

Let the circle with center O (Fig. 115) be the fundamental circle and AB any nominal line. We again wish to consider inversion with reference to the circle, center C, upon which AB lies. If PQ is any nominal segment, the circle upon which it lies will invert into another circle orthogonal to the fundamental circle, so that the nominal segment PQ will invert into a nominal segment $P'Q'$. Furthermore, if nominal line PQ meets the nominal line AB in a point M, the nominal line $P'Q'$ meets AB at the same point and makes the same angle with it. Moreover, the length of nominal segment PQ is equal to that of $P'Q'$, as we shall show, regardless of whether PQ and AB intersect or not.

Let the circle upon which the nominal line PQ lies meet the fundamental circle in points S and T, and that upon which the nominal line $P'Q'$ lies meet it in S' and T'. Then points S and S' are inverse points and so also are points T and T'. It is easy to show that

$$\frac{PT}{P'T'} = \frac{CT}{CP'},$$

$$\frac{P'S'}{PS} = \frac{CP'}{CS},$$

$$\frac{QT}{Q'T'} = \frac{CT}{CQ'}$$

and

$$\frac{Q'S'}{QS} = \frac{CQ'}{CS},$$

so that

$$\log \frac{PT}{PS}\Big/\frac{QT}{QS} = \log \frac{P'T'}{P'S'}\Big/\frac{Q'T'}{Q'S'}.$$

Thus we infer that nominal length remains invariant under such an inversion. Consequently, in the geometry of the nominal lines, we may define reflection about any nominal line as axis as inversion with regard to the circle upon which the nominal line lies. Displacement in this geometry may be accomplished by a series of these nominal reflections exactly as in Hyperbolic Geometry. Such displacement alters neither nominal length nor angle and thus leaves figures unchanged in nominal size and shape.

It follows that our definition of nominal length satisfies all of the requirements which must be met if the analogy to the idea of length in Hyperbolic Geometry is to be complete. With nominal length and displacement defined, every proposition of Hyperbolic Geometry will have its counterpart in our nominal geometry.

EXERCISES

1. Prove that there are no similar triangles in the geometry of the nominal lines, i.e., that two triangles with the three angles of one equal to the three angles of the other are congruent.

2. Regarding a diameter ST of the fundamental circle as a nominal line, prove that if P is a point on it such that its distance from the center O is unity, then $\dfrac{PS}{PT} = \epsilon$, where $k = 1$. This unit OP can be used for the purpose of comparison; every other unit segment can be made to coincide with it by displacement.

3. Extending the analogy between this nominal geometry and Hyperbolic Geometry, prove that the angles of parallelism for unequal distances are unequal and that the smaller angle corresponds to the greater distance.

4. Show how to construct on a diameter of the fundamental circle, regarded as a nominal line, the nominal distance corresponding to any given angle of parallelism.

103. The Counterparts of Circles, Limiting Curves and Equidistant Curves.

In Hyperbolic Geometry, a circle was regarded as the orthogonal trajectory of a sheaf of lines with an ordinary point for vertex. Since a sheaf of nominal lines with a nominal point for vertex consists of the system of coaxal circles passing through this point and its inverse with regard to the fundamental circle, the analogues of circles in the nominal geometry are the circles orthogonal to such a system and lying within the fundamental circle. Such circles are not orthogonal to the fundamental circle and hence are not nominal lines. They are *nominal circles* in the sense that each is the locus of points at a constant nominal distance from the vertex of a sheaf of nominal lines, this vertex being the *nominal center*. That these *nominal radii* are equal can be proved by moving the sheaf so that its vertex is at the center of the fundamental circle. As a consequence of this displacement, the system of coaxal circles becomes the sheaf of lines through the center and the orthogonal trajectories a system of concentric circles with nominal radii obviously equal.

Limiting curves are represented in this geometry by circles also. For a sheaf of parallel nominal lines consists of a coaxal system of circles of tangent type orthogonal to the fundamental circle, each tangent to all of the others at a point on the fundamental circle.

The orthogonal trajectories of such a sheaf are circles lying within and tangent to the fundamental circle at that point. They are not nominal lines, for they do not cut the fundamental circle at right angles. Nor are they nominal circles, for they are not closed. Each may be regarded as a circle with infinite radius, center at the *nominal ideal point* at which it touches the fundamental circle.

Finally, we consider a system of nominal lines all of which are perpendicular to a given nominal line. They will coincide with the

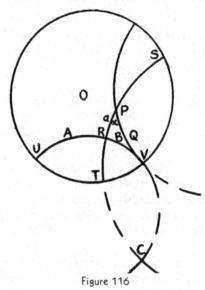

Figure 116

coaxal system of circles orthogonal to the fundamental circle and the circle on which the given line lies, having the points of intersection of those two circles as limiting points. The circles orthogonal to this coaxal system, that is, all circles through the two limiting points, are *nominal equidistant curves*. They are obviously neither nominal circles nor nominal limiting curves, for they do not lie entirely within the fundamental circle.

104. The Relation Between a Nominal Distance and its Angle of Parallelism.

By this time we have gone far enough with the development of the geometry of the nominal lines to convince the reader of the completeness of the resemblance to Hyperbolic Geometry. Further elaboration would be logically superfluous. Indeed, one does not need to do more than establish a one-to-one correspondence between the entities of the two geometries and recognize the analogy between the definitions and postulates of one and those of the other. But, in conclusion, it will prove interesting to derive for the nominal geometry the relation between a nominal distance. and its corresponding angle of parallelism.

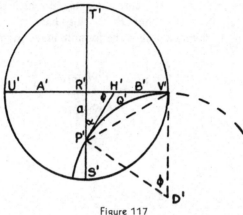

Figure 117

Choose the circle with center O (Fig. 116) as fundamental circle. Let P be any nominal point, AB any nominal line with coinciding circle cutting the fundamental circle at U and V, PQ one of the nominal lines parallel to AB through P with its coinciding circle touching that of AB at V, and PR the perpendicular nominal line from P to AB coinciding with a circle cutting the fundamental circle at S and T. Designate the nominal length of PR by a and the measure of the nominal angle of parallelism RPQ by α, the usual angular unit being implied.

Invert the figure with regard to the circle with center at C, the point of intersection outside the fundamental circle of the circles with which nominal lines AB and PR coincide, and orthogonal to the fundamental circle. This circle of inversion coincides with a nominal line and the inversion thus brings about a nominal displacement. The result of inversion. is shown in Figure 117. The fundamental circle remains as it was. The perpendicular nominal lines AB and PR invert into perpendicular lines $A'B'$ and $P'R'$ orthogonal to the inverse of the fundamental circle and hence passing through its center. The parallel PQ is transformed into a circle, center D', through P', orthogonal to the inverse of the fundamental circle and tangent to $A'B'$ at V'. This nominal reflection has left angles and lengths unchanged, hence, if $P'H'$ is the tangent line at P' to the circle with center D', angle $R'P'H'$ is equal to α and nominal segment $P'R'$ is equal to a. Designate by φ the measure of angle $R'H'P'$, H' being the point in which the tangent $P'H'$ cuts $A'B'$. Note that angle $P'D'V'$ also has measure φ.

Then, using for nominal length the formula involving the parameter,

$$a = k \log \frac{PT}{PS} \Big/ \frac{RT}{RS} = k \log \frac{P'T'}{P'S'} \Big/ \frac{R'T'}{R'S'}$$

$$= k \log \frac{P'T'}{P'S'}.$$

But, since

$$\varphi = \frac{\pi}{2} - \alpha$$

and

$$\angle R'V'P' = \frac{\varphi}{2},$$

we see that

$$\frac{P'R'}{r} = \tan\left(\frac{\pi}{4} - \frac{\alpha}{2}\right),$$

where r is the radius of the fundamental circle, so that

$$P'T' = r\left[1 + \tan\left(\frac{\pi}{4} - \frac{\alpha}{2}\right)\right]$$

and

$$P'S' = r\left[1 - \tan\left(\frac{\pi}{4} - \frac{\alpha}{2}\right)\right].$$

It follows that

$$a = k \log \cot \frac{\alpha}{2}$$

or

$$e^{-a/k} = \tan \frac{\alpha}{2},$$

a formula exactly like the corresponding one in Hyperbolic Geometry.

105. Conclusion.

There is no point to presenting here proof of the compatibility of the postulate system of Elliptic Geometry. Many such proofs exist. One, for example, much like the one above, compares Elliptic Geometry with a nominal geometry, essentially Euclidean, in which the nominal lines are circles intersecting a fixed circle in such a way that the chords which they have in common with that fundamental circle are its diameters.[8] But there are methods which are easier, although not based upon such elementary principles. The brief outline of the test for the consistency of Hyperbolic Geometry, just completed, suffices to reveal the nature and spirit of such proofs.

We conclude then that each of the three geometries is as consistent as either of the others. We are as thoroughly convinced that there is no contradiction to be encountered in Hyperbolic or Elliptic Geometry as we are that there is none in Euclidean. And we are as certain that Euclidean Geometry is consistent as it is possible to be about any body of reasoned doctrine.

It is clear now why all of the efforts to prove Euclid's Parallel Postulate were destined to end in failure. The Postulate can never be proved, for its proof would order the rejection of the parallel postulates of the equally consistent Non-Euclidean Geometries.

[8] See Carslaw: *Elements of Non-Euclidean Plane Geometry and Trigonometry*, pp. 171–174.

A P P E N D I X

I. THE FOUNDATION OF EUCLIDEAN GEOMETRY [1]

1. The Definitions of Book I.

1. A point is that which has no part.

2. A line is breadthless length.

3. The extremities of a line are points.

4. A straight line is a line which lies evenly with the points on itself.

5. A surface is that which has length and breadth only.

6. The extremities of a surface are lines.

7. A plane surface is a surface which lies evenly with the straight lines on itself.

8. A plane angle is the inclination to one another of two lines in a plane which meet one another and do not lie in a straight line.

9. And when the lines containing the angle are straight, the angle is called rectilineal.

10. When a straight line set up on a straight line makes the adjacent angles equal to one another, each of the equal angles is right, and the straight line standing on the other is called a perpendicular to that on which it stands.

11. An obtuse angle is an angle greater than a right angle.

12. An acute angle is an angle less than a right angle.

13. A boundary is that which is an extremity of anything.

14. A figure is that which is contained by any boundary or boundaries.

[1] From *The Thirteen Books of Euclid's Elements*, a translation from the text of Heiberg, with introduction and commentary, by Thomas L. Heath. By permission of The Macmillan Company, representing the Cambridge University Press.

15. A circle is a plane figure contained by one line such that all the straight lines falling upon it from one point among those lying within the figure are equal to one another.

16. And the point is called the centre of the circle.

17. A diameter of the circle is any straight line drawn through the centre and terminated in both directions by the circumference of the circle, and such a straight line also bisects the circle.

18. A semicircle is the figure contained by the diameter and the circumference cut off by it. And the centre of the semicircle is the same as that of the circle.

19. Rectilineal figures are those which are contained by straight lines, trilateral figures being those contained by three, quadrilateral those contained by four, and multilateral those contained by more than four straight lines.

20. Of trilateral figures, an equilateral triangle is that which has its three sides equal, an isosceles triangle that which has two of its sides alone equal, and a scalene triangle that which has its three sides unequal.

21. Further, of trilateral figures, a right-angled triangle is that which has a right angle, an obtuse-angled triangle that which has an obtuse angle, and an acute-angled triangle that which has its three angles acute.

22. Of quadrilateral figures, a square is that which is both equilateral and right-angled; an oblong that which is right-angled but not equilateral; a rhombus that which is equilateral but not right-angled; and a rhomboid that which has its opposite sides and angles equal to one another but is neither equilateral nor right-angled. And let quadrilaterals other than these be called trapezia.

23. Parallel straight lines are straight lines which, being in the same plane and being produced indefinitely in both directions, do not meet one another in either direction.

2. The Postulates.

Let the following be postulated:

1. To draw a straight line from any point to any point.

2. To produce a finite straight line continuously in a straight line.

3. To describe a circle with any centre and distance.

4. That all right angles are equal to one another.

5. That, if a straight line falling on two straight lines make the interior angles on the same side less than two right angles, the two straight lines, if produced indefinitely, meet on that side on which are the angles less than the two right angles.

3. The Common Notions.

1. Things which are equal to the same thing are also equal to one another.

2. If equals be added to equals, the wholes are equal.

3. If equals be subtracted from equals, the remainders are equal.

4. Things which coincide with one another are equal to one another.

5. The whole is greater than the part.

4. The Forty-Eight Propositions of Book I.

1. On a given finite straight line to construct an equilateral triangle.

2. To place at a given point (as an extremity) a straight line equal to a given straight line.

3. Given two unequal straight lines, to cut off from the greater a straight line equal to the less.

4. If two triangles have the two sides equal to two sides respectively, and have the angles contained by the equal straight lines equal, they will also have the base equal to the base, the triangle will be equal to the triangle, and the remaining angles will be equal to the remaining angles respectively, namely those which the equal sides subtend.

5. In isosceles triangles the angles at the base are equal to one another, and, if the equal straight lines be produced further, the angles under the base will be equal to one another.

6. If in a triangle two angles be equal to one another, the sides which subtend the equal angles will also be equal to one another.

7. Given two straight lines constructed on a straight line (from its extremities) and meeting in a point, there cannot be constructed on the same straight line (from its extremities), and on the same side of it, two other straight lines meeting in another point and equal to the former two respectively, namely each to that which has the same extremity with it.

8. If two triangles have the two sides equal to two sides respectively, and have also the base equal to the base, they will also have the angles equal which are contained by the equal straight lines.

9. To bisect a given rectilineal angle.

10. To bisect a given finite straight line.

11. To draw a straight line at right angles to a given straight line from a given point on it.

12. To a given infinite straight line, from a given point which is not on it, to draw a perpendicular straight line.

13. If a straight line set up on a straight line make angles, it will make either two right angles or angles equal to two right angles.

14. If with any straight line, and at a point on it, two straight lines not lying on the same side make the adjacent angles equal to two right angles, the two straight lines will be in a straight line with one another.

15. If two straight lines cut one another, they make the vertical angles equal to one another.

16. In any triangle, if one of the sides be produced, the exterior angle is greater than either of the interior and opposite angles.

17. In any triangle two angles taken together in any manner are less than two right angles.

18. In any triangle the greater side subtends the greater angle.

19. In any triangle the greater angle is subtended by the greater side.

20. In any triangle two sides taken together in any manner are greater than the remaining one.

21. If on one of the sides of a triangle, from its extremities, there be constructed two straight lines meeting within the triangle, the straight lines so constructed will be less than the remaining two sides of the triangle, but will contain a greater angle.

22. Out of three straight lines, which are equal to three given straight lines, to construct a triangle: thus it is necessary that two of the straight lines taken together in any manner should be greater than the remaining one.

23. On a given straight line and at a point on it to construct a rectilineal angle equal to a given rectilineal angle.

24. If two triangles have the two sides equal to two sides re-

spectively, but have the one of the angles contained by the equal straight lines greater than the other, they will also have the base greater than the base.

25. If two triangles have the two sides equal to two sides respectively, but have the base greater than the base, they will also have the one of the angles contained by the equal straight lines greater than the other.

26. If two triangles have the two angles equal to two angles respectively, and one side equal to one side, namely, either the side adjoining the equal angles, or that subtending one of the equal angles, they will also have the remaining sides equal to the remaining sides and the remaining angle to the remaining angle.

27. If a straight line falling on two straight lines make the alternate angles equal to one another, the straight lines will be parallel to one another.

28. If a straight line falling on two straight lines make the exterior angle equal to the interior and opposite angle on the same side, or the interior angles on the same side equal to two right angles, the straight lines will be parallel to one another.

29. A straight line falling on parallel straight lines makes the alternate angles equal to one another, the exterior angle equal to the interior and opposite angle, and the interior angles on the same side equal to two right angles.

30. Straight lines parallel to the same straight line are also parallel to one another.

31. Through a given point to draw a straight line parallel to a given straight line.

32. In any triangle, if one of the sides be produced, the exterior angle is equal to the two interior and opposite angles, and the three interior angles of the triangle are equal to two right angles.

33. The straight lines joining equal and parallel straight lines (at the extremities which are) in the same directions (respectively) are themselves also equal and parallel.

34. In parallelogrammic areas the opposite sides and angles are equal to one another, and the diameter bisects the areas.

35. Parallelograms which are on the same base and in the same parallels are equal to one another.

36. Parallelograms which are on equal bases and in the same parallels are equal to one another.

37. Triangles which are on the same base and in the same parallels are equal to one another.

38. Triangles which are on equal bases and in the same parallels are equal to one another.

39. Equal triangles which are on the same base and on the same side are also in the same parallels.

40. Equal triangles which are on equal bases and on the same side are also in the same parallels.

41. If a parallelogram have the same base with a triangle and be in the same parallels, the parallelogram is double of the triangle.

42. To construct, in a given rectilineal angle, a parallelogram equal to a given triangle.

43. In any parallelogram the complements of the parallelograms about the diameter are equal to one another.

44. To a given straight line to apply, in a given rectilineal angle, a parallelogram equal to a given triangle.

45. To construct, in a given rectilineal angle, a parallelogram equal to a given rectilineal figure.

46. On a given straight line to describe a square.

47. In right-angled triangles the square on the side subtending the right angle is equal to the squares on the sides containing the right angle.

48. If in a triangle the square on one of the sides be equal to the squares on the remaining two sides of the triangle, the angle contained by the remaining two sides of the triangle is right.

II. CIRCULAR AND HYPERBOLIC FUNCTIONS

5. The Trigonometric Functions.

It is assumed that the student is acquainted with the following infinite power series, the Maclaurin expansions for the exponential function e^x and the trigonometric or, as they are frequently called, circular functions $\sin x$ and $\cos x$:

$$e^x = 1 + x + \frac{x^2}{2!} + \frac{x^3}{3!} + \dots + \frac{x^{n-1}}{(n-1)!} + \dots,$$

$$\sin x = x - \frac{x^3}{3!} + \frac{x^5}{5!} - \dots + (-1)^{n+1}\frac{x^{2n-1}}{(2n-1)!} + \dots,$$

$$\cos x = 1 - \frac{x^2}{2!} + \frac{x^4}{4!} - \dots + (-1)^{n+1}\frac{x^{2n-2}}{(2n-2)!} + \dots$$

It will be recalled that these series are convergent and that each defines a continuous function of x for all real values of x. We regard these series as the definitions of the functions e^x, $\sin x$ and $\cos x$. This generalization relieves us of the restrictions imposed by the special definitions of elementary trigonometry. For example, $\sin x$ no longer necessarily refers to a certain ratio of two sides of a right triangle with acute angle x any more than the function x^2 necessarily represents the area of a square with side x. Indeed, for the functions $\sin x$ and $\cos x$, x is, from the general viewpoint, to be regarded as an abstract number and not as an angle at all. When it is regarded as the measure of an angle, there result merely special applications of our definitions.

It is known that these series may be added to and subtracted from each other, multiplied and divided, one by another, and the resulting series will converge for all values of x also, excepting those for which in division the divisor converges to zero. They may also be differentiated or integrated term by term. Thus we can easily obtain the infinite power series representations for the other trigonometric functions, for example,

$$\tan x = \frac{\sin x}{\cos x} = x + \frac{x^3}{3} + \frac{2x^5}{15} + \cdots$$

From these general definitions, all of the well-known properties of the trigonometric functions can be derived, such as

$$\sin^2 x + \cos^2 x = 1,$$
$$\sin 2x = 2 \sin x \cos x,$$
$$\cos x \cos y - \sin x \sin y = \cos (x + y),$$
$$\cos x + \cos y = 2 \cos \frac{x + y}{2} \cos \frac{x - y}{2},$$
$$\frac{d}{dx} \sin x = \cos x$$

and, where $\sec x$ is defined as $\dfrac{1}{\cos x}$,

$$\sec^2 x - \tan^2 x = 1.$$

This new freedom allows us to extend our ideas of e^x, $\sin x$, $\cos x$ and $\tan x$ even to the cases where x is an imaginary number. Thus if $x = a + bi$, where a and b are real and $i = \sqrt{-1}$, we define, for example, $\sin (a + bi)$ as follows:

$$\sin (a + bi) = (a + bi) - \frac{(a + bi)^3}{3!} + \frac{(a + bi)^5}{5!} - \cdots$$

From this broader viewpoint we recognize that the trigonometric identities and relationships hold for complex and not only real arguments.

Of special interest are the cases in which x is a pure imaginary. We have in particular

$$e^{xi} = 1 + xi + \frac{(xi)^2}{2!} + \frac{(xi)^3}{3!} + \frac{(xi)^4}{4!} + \frac{(xi)^5}{5!} + \cdots$$

$$= \left(1 - \frac{x^2}{2!} + \frac{x^4}{4!} - \cdots\right) + \left(x - \frac{x^3}{3!} + \frac{x^5}{5!} - \cdots\right)i$$

or

$$e^{xi} = \cos x + i \sin x.$$

Similarly

$$e^{-xi} = \cos x - i \sin x.$$

These last two results yield the remarkable formulas,

$$\sin x = \frac{e^{xi} - e^{-xi}}{2i},$$

$$\cos x = \frac{e^{xi} + e^{-xi}}{2}.$$

These formulas may be used as alternative definitions of $\sin x$ and $\cos x$. Starting with these, all of the familiar formulas and relations connecting the trigonometric functions can be derived.

6. The Hyperbolic Functions.

The last two formulas of the preceding section suggest two new functions of x which are called the *hyperbolic sine of x* and the *hyperbolic cosine of x* and are defined thus:

$$\sinh x = \frac{e^x - e^{-x}}{2},$$

$$\cosh x = \frac{e^x + e^{-x}}{2}.$$

Related to these are four other hyperbolic functions, *hyperbolic tangent, secant, cosecant* and *cotangent*, defined as follows:

$$\tanh x = \frac{\sinh x}{\cosh x} = \frac{e^x - e^{-x}}{e^x + e^{-x}} = \frac{e^{2x} - 1}{e^{2x} + 1},$$

$$\text{sech } x = \frac{1}{\cosh x},$$

$$\operatorname{csch}\,x = \frac{1}{\sinh x},$$

$$\coth x = \frac{1}{\tanh x}.$$

Reversing the procedure of the last section, we can obtain power series definitions for the hyperbolic functions by substitution of the power series for e^x and e^{-x} in the formulas defining sinh x and cosh x above. We obtain

$$\sinh x = x + \frac{x^3}{3!} + \frac{x^5}{5!} + \frac{x^7}{7!} + \ldots,$$

$$\cosh x = 1 + \frac{x^2}{2!} + \frac{x^4}{4!} + \frac{x^6}{6!} + \ldots.$$

The similarity of these series to those for sin x and cos x suggests a simple relationship between the circular and hyperbolic functions. By replacing x by xi in the power series for sin x and cos x, we obtain

$$\sin xi = i \sinh x,$$

$$\cos xi = \cosh x,$$

from which follow

$$\tan xi = i \tanh x,$$

$$\csc xi = -i \operatorname{csch} x,$$

$$\sec xi = \operatorname{sech} x,$$

$$\cot xi = -i \coth x.$$

Thus the hyperbolic functions can be defined in terms of the exponential function, by infinite power series, and in terms of the circular functions. From any one of these viewpoints can be obtained the formulas and relations connecting the hyperbolic functions analogous to those for circular functions. Thus, starting with the familiar identity

$$\sin^2 x + \cos^2 x = 1,$$

we have, on replacing x by xi,

$$\sin^2 xi + \cos^2 xi = 1$$

or

$$(i \sinh x)^2 + \cosh^2 x = 1,$$

and finally

$$\cosh^2 x - \sinh^2 x = 1.$$

On the other hand this relation can be verified by substituting the exponential forms of the hyperbolic functions, thus

$$\left(\frac{e^x + e^{-x}}{2}\right)^2 - \left(\frac{e^x - e^{-x}}{2}\right)^2 = 1,$$

or by use of the series expansions:

$$\left(1 + \frac{x^2}{2!} + \frac{x^4}{4!} + \frac{x^6}{6!} + \ldots\right)^2 - \left(x + \frac{x^3}{3!} + \frac{x^5}{5!} + \ldots\right)^2 = 1.$$

Other relations will be found in the list of exercises.

The formulas for the differentiation of the hyperbolic functions are readily obtained. For example, by differentiating $\dfrac{e^x - e^{-x}}{2}$, or

the series $x + \dfrac{x^3}{3!} + \dfrac{x^5}{5!} + \ldots$, or $\dfrac{\sin xi}{i}$, one obtains the result

$$\frac{d}{dx} \sinh x = \cosh x.$$

In Figure 118 are shown the graphs of the functions sinh x, cosh x and tanh x. The hyperbolic cosine curve is the familiar *catenary*.

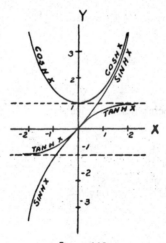

Figure 118

An important limit,

$$\lim_{u \to 0} \frac{\sinh u}{u} = 1,$$

can be verified in a variety of ways. For example, the method of differential calculus for evaluating so-called indeterminate forms may be used.   Otherwise, sinh u may be replaced by the infinite series or in terms of the exponential function, or $\dfrac{\sinh u}{u}$ may be written as $\dfrac{\sin iu}{iu}$, before the limit is taken.

7. The Inverse Hyperbolic Functions.

If $y = \sinh x$, then x is defined as a function of y.   We say that x is the *inverse hyperbolic sine* of y and write

$$x = \sinh^{-1} y.$$

Similarly there is an inverse function corresponding to each of the other hyperbolic functions.

Since the hyperbolic functions are expressible in terms of the exponential function, it is to be expected that the inverse hyperbolic functions can be represented in terms of logarithms.   For example, if $y = \sinh^{-1} x$, we have

$$x = \sinh y = \frac{e^y - e^{-y}}{2},$$

or

$$e^{2y} - 2xe^y - 1 = 0.$$

Then

$$e^y = x + \sqrt{x^2 + 1},$$

the negative sign being deleted since e^y is positive when y is real, so that

$$\sinh^{-1} x = \log (x + \sqrt{x^2 + 1}).$$

The expressions for the inverse hyperbolic cosine and tangent will be found in the list of exercises.

Again, if $y = \sinh^{-1} x$, we have

$$x = \sinh y$$

and

$$\frac{dx}{dy} = \cosh y.$$

Whence

$$\frac{dy}{dx} = \frac{1}{\cosh y} = \frac{1}{\sqrt{1 + \sinh^2 y}},$$

the positive sign being used with the radical since cosh y is always positive when y is real. Thus

$$\frac{d}{dx}\sinh^{-1} x = \frac{1}{\sqrt{1 + x^2}}.$$

This result may be obtained otherwise by differentiating the expression for $\sinh^{-1} x$ in terms of x derived above. Other differentiation formulas will be found below.

8. Geometric Interpretation of Circular and Hyperbolic Functions.

The equations $x = \cos z$, $y = \sin z$ may be regarded as parametric equations of the unit circle $x^2 + y^2 = 1$. The parameter z is generally interpreted geometrically as the measure of the central angle POQ (Fig. 119). It will be instructive in connection with our study

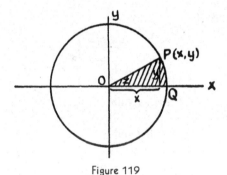

Figure 119

of hyperbolic functions to interpret this parameter in another way, namely, as twice the measure of the area of the circular sector swept out by the radius OP as P moves along the circle from Q to any point (x, y). The validity of this interpretation is easily verified since

$$\text{Area of Sector} = \frac{z}{2\pi} \cdot \pi = \frac{z}{2}.$$

Turning now to the hyperbolic functions we propose by way of comparison to regard the equations $x = \cosh z$, $y = \sinh z$ as parametric equations of the equilateral hyperbola $x^2 - y^2 = 1$ and show that the parameter z can be interpreted as twice the measure of the area of the hyperbolic sector swept out by the vector OP (Fig. 120) as P moves along the hyperbola from Q to any point (x, y).

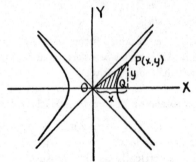

Figure 120

$$\text{Area of Sector} = \frac{xy}{2} - \int_{1}^{x} y\,dx$$

$$= \frac{\cosh z \sinh z}{2} - \int_{0}^{z} \sinh^{2} z\,dz$$

$$= \frac{\cosh z \sinh z}{2} - \int_{0}^{z} \frac{\cosh 2z - 1}{2}\,dz$$

$$= \frac{\cosh z \sinh z}{2} - \left[\frac{\sinh 2z}{4} - \frac{z}{2}\right]_{0}^{z}$$

$$= \frac{\cosh z \sinh z}{2} - \frac{\cosh z \sinh z}{2} + \frac{z}{2} = \frac{z}{2}.$$

EXERCISES

Verify the following:

1. (a) $\operatorname{sech}^2 x + \tanh^2 x = 1$,
   (b) $\coth^2 x - \operatorname{csch}^2 x = 1$,
   (c) $\sinh x + \cosh x = e^x$.

2. (a) $\sinh (x \pm y) = \sinh x \cosh y \pm \cosh x \sinh y$,
   (b) $\cosh (x \pm y) = \cosh x \cosh y \pm \sinh x \sinh y$,
   (c) $\tanh (x \pm y) = \dfrac{\tanh x \pm \tanh y}{1 \pm \tanh x \tanh y}$.

3. (a) $\sinh 2x = 2 \sinh x \cosh x$,
   (b) $\cosh 2x = \sinh^2 x + \cosh^2 x = 1 + 2 \sinh^2 x = 2 \cosh^2 x - 1$,
   (c) $\tanh 2x = \dfrac{2 \tanh x}{1 + \tanh^2 x}$.

4. (a) $\sinh \dfrac{x}{2} = \pm \sqrt{\dfrac{\cosh x - 1}{2}}$,

(b) $\cosh \dfrac{x}{2} = \sqrt{\dfrac{\cosh x + 1}{2}}$,

(c) $\tanh \dfrac{x}{2} = \pm \sqrt{\dfrac{\cosh x - 1}{\cosh x + 1}}$.

5. (a) $\sinh x + \sinh y = 2 \sinh \dfrac{x+y}{2} \cosh \dfrac{x-y}{2}$,

   (b) $\sinh x - \sinh y = 2 \cosh \dfrac{x+y}{2} \sinh \dfrac{x-y}{2}$,

   (c) $\cosh x + \cosh y = 2 \cosh \dfrac{x+y}{2} \cosh \dfrac{x-y}{2}$,

   (d) $\cosh x - \cosh y = 2 \sinh \dfrac{x+y}{2} \sinh \dfrac{x-y}{2}$.

6. (a) $\sinh(-x) = -\sinh x$,
   (b) $\cosh(-x) = \cosh x$,
   (c) $\tanh(-x) = -\tanh x$.

7. (a) $\dfrac{d}{dx} \cosh x = \sinh x$,

   (b) $\dfrac{d}{dx} \tanh x = \operatorname{sech}^2 x$,

   (c) $\dfrac{d}{dx} \coth x = -\operatorname{csch}^2 x$,

   (d) $\dfrac{d}{dx} \operatorname{sech} x = -\operatorname{sech} x \tanh x$,

   (e) $\dfrac{d}{dx} \operatorname{csch} x = -\operatorname{csch} x \coth x$.

8. $\lim\limits_{u \to 0} \dfrac{\tanh u}{u} = 1$.

9. $\tanh x = x - \dfrac{x^3}{3} + \dfrac{2x^5}{15} - \dfrac{17x^7}{315} + \cdots$

10. (a) $\cosh^{-1} x = \pm \log(x + \sqrt{x^2 - 1})$,

   (b) $\tanh^{-1} x = \dfrac{1}{2} \log \dfrac{1+x}{1-x}$.

11. (a) $\dfrac{d}{dx} \cosh^{-1} x = \pm \dfrac{1}{\sqrt{x^2 - 1}}$,

   (b) $\dfrac{d}{dx} \tanh^{-1} x = \dfrac{1}{1 - x^2}$.

12. (a) $\displaystyle\int \dfrac{dx}{\sqrt{x^2 + a^2}} = \sinh^{-1} \dfrac{x}{a} + C$,

   (b) $\displaystyle\int \dfrac{dx}{\sqrt{x^2 - a^2}} = \cosh^{-1} \dfrac{x}{a} + C$,

   (c) $\displaystyle\int \dfrac{dx}{a^2 - x^2} = \dfrac{1}{a} \tanh^{-1} \dfrac{x}{a} + C$.

13. (a) $\sinh 0 = 0$,

    (b) $\cosh 0 = 1$,

    (c) $\sinh \dfrac{1}{2} = .521$,

    (d) $\lim\limits_{x \to \infty} \tanh x = 1$,

    (e) $\cosh \dfrac{1}{5} = ?$

III. THE THEORY OF ORTHOGONAL CIRCLES AND ALLIED TOPICS

9. The Power of a Point with Regard to a Circle.

It is well known that if, through a point P in the plane of a circle, a secant line is drawn cutting the circle in points A and B, then the product $PA \cdot PB$ is constant. This product is positive, negative or zero, accordingly as P is outside, inside or on the circle. When P lies outside, the product is the square of the tangential distance from the point to the circle. This constant product we call the *power* of the point with regard to the circle. Then, if O is the center of the circle and r the radius,

$$\text{Power} = (PO + r)(PO - r) = \overline{PO}^2 - r^2.$$

10. The Radical Axis of Two Circles.

Theorem 1. If a point moves in such a way that its power with regard to one of two circles is always equal to its power with regard to the other, its locus is a straight line perpendicular to the line of centers of the two circles.

Let P (Fig. 121) be any point which has equal powers with regard to the two circles with centers at O_1 and O_2 and radii r_1 and r_2. Then, if PQ is drawn perpendicular to O_1O_2, we have

$$\overline{PO_1}^2 - \overline{QO_1}^2 = \overline{PO_2}^2 - \overline{QO_2}^2$$

and

$$\overline{QO_1}^2 - \overline{QO_2}^2 = (\overline{PO_1}^2 - r_1^2) - (\overline{PO_2}^2 - r_2^2) + r_1^2 - r_2^2,$$

or

$$(QO_1 + QO_2)(QO_1 - QO_2) = r_1^2 - r_2^2.$$

Since the product of the factors on the left is constant and one of them is equal to O_1O_2, the other is constant and hence so also is their sum and difference. It follows that as P moves so that its powers with regard to the two circles remain equal, the foot of the perpendicular from it to O_1O_2 is fixed, and thus that P lies on a straight line perpendicular to O_1O_2. It is easy to show that every point on this line has equal powers with regard to the two circles.

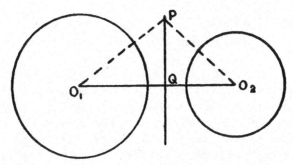

Figure 121

The line which is the locus of points whose powers with regard to two circles are equal is called the *radical axis* of the two circles.

Corollary. If two circles intersect, their radical axis is the line joining their points of intersection. If they are tangent to one another, their radical axis is the common tangent at the point of tangency.

Theorem 2. The three radical axes of three circles taken in pairs are in general concurrent.

The proof is left to the reader.

The point of intersection of the three radical axes of three circles taken in pairs is called the *radical center* of the three circles. It has the property that its powers with regard to the three circles are equal. Theorem 2 provides an easy method of constructing the radical axis of two given circles which do not intersect. To obtain a point on the radical axis, one need only construct any circle which intersects both of the given circles; the two radical axes thus determined intersect in general on the required radical axis.

A point may be regarded as a circle with radius zero. The results obtained above hold for *point circles* as well as circles proper. Thus the power of a point with regard to a point circle is the square of the distance between the points; the radical axis of a circle and a point on its circumference regarded as a point circle is the tangent to the circle at the point; the radical axis of two point circles is the perpendicular bisector of the segment joining them.

11. Orthogonal Circles.

When two circles intersect one another in such a way that their tangent lines at a point of intersection are perpendicular, they are said to cut *orthogonally* and each is *orthogonal* to the other. As a consequence of symmetry, the tangent lines at the second point of cutting will, under these circumstances, also be perpendicular. It follows from the definition that two circles are orthogonal if and only if the tangent lines of each at the points of intersection pass through the center of the other. Thus the center of each of two orthogonal circles must lie outside the other.

Theorem. If two circles are orthogonal, the square of the radius of each is the power of its center with regard to the other. Conversely, if the square of the radius of one circle is the power of its center with regard to another, the two circles are orthogonal.

The reader may supply the proof.

Thus it appears that a circle can be constructed orthogonal to a given circle with center at any given point outside the given circle; its radius will be the tangential distance from the point to the circle. In order to be orthogonal to two given circles, a circle must have its center on the radical axis of the given circles and outside those circles if they intersect.

If three given circles have, as is generally the case, a single radical center, and if it lies outside the circles, one and only one circle can be constructed orthogonal to all three circles; its center is the radical center and its radius is the common tangential distance from the center to the circles.

Any point lying on the circumference of a given circle may be regarded as a point circle orthogonal to the given circle. As a consequence, we recognize that an infinite number of circles can be drawn through a given point and orthogonal to a given circle, provided the point is not the center of the circle. But only one circle, in general, can be drawn through a point orthogonal to two given circles, exceptions occurring when the point is collinear with the centers of the circles. Under the latter circumstance, no such circle can be constructed, unless the two circles are tangent and the given point is the point of tangency, when an infinite number of such circles can be drawn. Through two given points can be drawn, in general, only one circle orthogonal to a given circle. This is always the case if the points are not collinear with the center of the given circle. When the given circle has its center collinear with the given points, no circle can be constructed orthogonal to it and passing through the given points, unless the points lie on the same side of the center in such a way that the product of the distances from the center to the two points is equal to the square of the radius of the circle. In the latter case, *all* circles through the two points are orthogonal to the circle, because the power of the center of the latter with regard to every circle through the two given points is equal to the square of its radius.

12. Systems of Coaxal Circles.

A system of circles such that the radical axis of any two of them is the same as the radical axis of any other pair is called a *coaxal system*. It should be clear, as a consequence of the definition, that the centers of the circles of a coaxal system lie on a line perpendicular to the common radical axis, that every point of this radical axis has equal powers with regard to all of the circles of the system, and that a circle orthogonal to any two circles of the system is orthogonal to all of them. Two circles of a coaxal system determine the system; if two of the circles are given, any other can be constructed.

There are three types of coaxal system: *intersecting, tangent*, and *non-intersecting*. If two circles of a system intersect in points *A* and

B (Fig. 122), all of the circles pass through those two points. The system is called an *intersecting system*. If the system of circles is constructed, each of which is orthogonal to all of the circles of this intersecting coaxal system, there results another coaxal system.

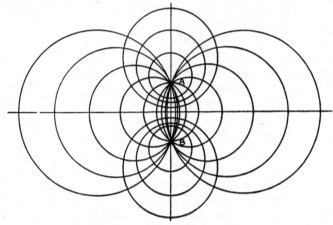

Figure 122

For every circle of the first system is orthogonal to every pair of circles of the new one, making the line of centers of the first a common radical axis for the pairs of the second.

The new system is of the *non-intersecting* type, since no circle orthogonal to two intersecting circles can intersect the line of centers of those circles. The two points of intersection of the intersecting system, when regarded as point circles, belong to the non-intersecting system. They are called the *limiting points* of the latter system. All circles orthogonal to the circles of a non-intersecting coaxal system pass through its limiting points and constitute an intersecting coaxal system.

The system of all circles tangent to a straight line at the same point is a coaxal system of the *tangent* type. The circles orthogonal to all of the circles of such a system form another coaxal system of the same type.

EXERCISES

1. Under what condition will two circles have no radical axis? When will three circles have no radical center? When will three circles have an infinite number of radical centers?

2. Construct a circle passing through a given point and orthogonal to a given circle.

3. Construct the circle passing through two given points and orthogonal to a given circle. Under what circumstances is the construction impossible? When can more than one such circle be drawn?

4. Construct the circle through a given point and orthogonal to two given circles.

5. Show that if two circles intersect and each is orthogonal to a third circle, then one point of intersection lies inside, the other outside, the third circle.

6. If two points C and D divide a diameter AB of a circle, center O, internally and externally in the same ratio, prove that

$$\overline{OA}^2 = \overline{OB}^2 = OC \cdot OD.$$

7. Prove that a circle orthogonal to two given circles will intersect, be tangent to or not intersect their line of centers accordingly as they do not intersect, are tangent to or intersect one another, respectively.

IV. THE ELEMENTS OF INVERSION

13. Inversion.

Choose any point P in the plane of a fixed circle with center at O and radius r. On OP construct the point P' such that $OP \cdot OP' = r^2$. Points P and P' are called *inverse points* and, since the relationship is mutual, each is called the *inverse* of the other. The fixed circle is called the *circle of inversion*, O the *center of inversion* and r^2 the *constant of inversion*. Thus through the medium of a circle there is set up a one-to-one correspondence between the points of a plane; to every point, with the exception of the center of inversion, there is a corresponding point.

Regarding inversion as a transformation of the plane into itself, it appears that points inside the circle of inversion are transformed into points outside and *vice versa*. The points on the circle of inversion are fixed. If a moving point traces any curve continuously, its inverse will trace continuously a curve called the *inverse* of the first. If a curve intersects the circle of inversion, its inverse will intersect it in the same point. The circle of inversion is absolutely fixed. Straight lines through the center of inversion are also fixed under inversion, although there is a redistribution of the points. It should be clear, from what has been said before, that every circle orthogonal to the circle of inversion is fixed, for the power of the center of inversion with regard to such a circle is r^2.

Theorem. If A, A′ and B, B′ are any two pairs of inverse points which do not lie on the same diameter of the circle of inversion, then they lie on a circle and angles OAB and OBA are equal, respectively, to angles OB′A′ and OA′B′.

If A, A' and B, B' (Fig. 123)[2] are pairs of inverse points, and 0 is the center of inversion, then

$$OA \cdot OA' = OB \cdot OB',$$

which implies that points A, A', B, B' lie on a circle. It is then easy to see that angles OAB and $OB'A'$ are equal and also angles OBA and $OA'B'$.

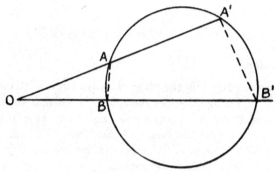

Figure 123

14. The Inverse of a Circle and the Inverse of a Line.

While straight lines through the center of inversion invert into themselves, lines in general do not invert into lines. We have the following theorem:

Theorem 1. Every straight line which does not pass through the center of inversion inverts into a circle passing through the center of inversion, and conversely.

Let l (Fig. 124) be any straight line not passing through the center of inversion 0. Draw from 0 the perpendicular OA to l. Let A' be the inverse of A and B' the inverse of B, any other point on l.

---

[2] Frequently one does not take the trouble to draw the circle of inversion.

Then, since angle OAB is a right angle, so also is angle $OB'A'$. Hence, as B moves along l, B' traces a circle with OA' as diameter. The proof is easily reversed.

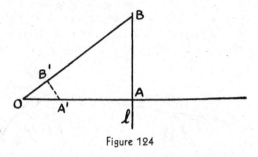

Figure 124

Theorem 2. Circles which do not pass through the center of inversion invert into circles.

Let C (Fig. 125) be the center of any circle not passing through the center of inversion 0. Draw OC, cutting the circle in points Q

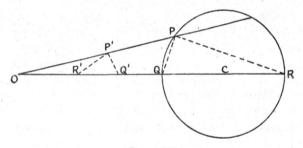

Figure 125

and R. Let P be any other point on the circle. Designate by P', Q', R' the points which are the inverses of P, Q, R, respectively. Then, since angles $OP'Q'$ and OQP are equal, as well as angles $OP'R'$ and ORP, it follows that angle $R'P'Q'$ is equal to angle QPR. But the latter angle is always a right angle, as P traces the circle. Then angle $R'P'Q'$ is always a right angle and P' traces a circle with the points R' and Q' as the extremities of a diameter. It is to be observed that the center of this circle is not the inverse of C.

15. The Effect of Inversion on Angles.

Let us turn our attention to any two curves PR and QR (Fig. 126) intersecting at R, the points P and Q being collinear with O, the center of inversion. Then the inverse curves $P'R'$ and $Q'R'$ will intersect in R', the inverse of R. The reader will have no difficulty in showing that angles PRQ and $Q'R'P'$ are equal. Now, as P

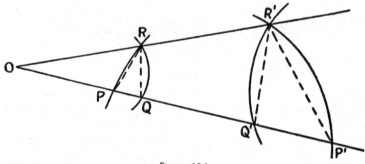

Figure 126

moves along curve PR continuously and approaches R, the secant PR approaches as a limiting position the tangent line to curve PR at R. At the same time Q approaches R and the secant QR approaches the tangent line to curve QR at R, while secants $P'R'$ and $Q'R'$ approach the tangents to the inverse curves at R'. Since angles PRQ and $Q'R'P'$ are always equal, as P approaches R, they are equal in the limit. We thus discover that the angle between two curves is the same as the angle between the inverse curves. In other words, inversion preserves angles. Such a transformation is referred to as *conformal*.

Theorem. Inversion is a conformal transformation.

---

16. The Peaucellier Inversor.

Although it has nothing in particular to do with our investigations, we must not conclude even a brief *résumé* of the theory of inversion without directing attention to a beautiful instrument for accomplishing inversion mechanically. This device, known as the *Peaucellier Inversor*, is a linkage consisting of six rigid bars. Four of these, *AB*, *BC*, *CD* and *DA* in Figure 127, are of equal length and hinged to one another to form a rhombus. The other equal bars, *OB* and *OD*, are hinged together at *O*, their other extremities being joined to the rhombus at *B* and *D*. It is easy to see that *O*, *A* and *C* are always collinear. Furthermore, if the circle with center at *D* and passing through *A* and *C* is constructed, it is clear that

$$OA \cdot OC = (OD - DA)(OD + DA) = \overline{OD}^2 - \overline{DA}^2,$$

so that *C* is the inverse of *A*, with *O* as center of inversion and $\overline{OD}^2 - \overline{DA}^2$ as constant of inversion. Then if *O* is fixed in the plane and *A* allowed to trace a curve, *C* will trace the inverse curve.

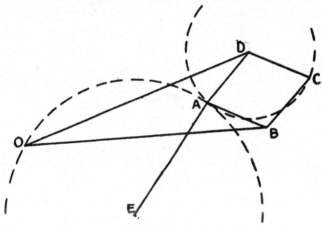

Figure 127

Perhaps the most striking application of the device is its employment in the conversion of circular motion into straight-line or, as it is frequently called, *rectilinear* motion. If, by the addition of a seventh link, a radial bar hinged to the rhombus at *A* and to a fixed point *E* of the plane, the point *A* is constrained to move along a circle through the center of inversion *O*, then *C* will travel along a straight line.

Euclid postulated the ability to construct a circle with any center and any radius and to construct straight lines. The reader is acquainted with a simple instrument, theoretically perfect, for the construction of circles. Here he encounters, perhaps for the first time, an instrument for the construction of straight lines. The straight-edge, which he has been accustomed to use, is but a pattern to be traced along, and implies some previous construction of a straight line.

Peaucellier, a captain in the French Army, proposed in a communication, published in 1864 in the Nouvelles Annales (Second Series, Vol. III, pp. 414–15), the problem of converting circular into rectilinear motion by means of a linkage, indicating that he himself had a solution. His communication attracted little or no attention. Indeed, when his device was rediscovered by a Russian named Lipkin, the latter was given the credit for the discovery. The matter was rectified later, however. Peaucellier's solution was published in 1873. Since then many other inversors have been devised, some with only four bars. Thus it is possible to convert circular into rectilinear motion with a linkwork consisting of five bars instead of the seven used in the original instrument.

EXERCISES

1. If A and A' are inverse points, prove that any circle through them is orthogonal to the circle of inversion.

2. If A, A' and B, B' are pairs of inverse points, O the center and r^2 the constant of inversion, show that

$$AB = \frac{r^2}{OA'OB'} A'B'.$$

3. If two circles are orthogonal, show that the inverse of the center of the first, with respect to the second as circle of inversion, coincides with the inverse of the center of the second with respect to the first.

4. Show how to choose the circle of inversion so that each of three given circles will invert into itself. Under what circumstances will this be impossible?

5. What is the inverse of an intersecting system of coaxal circles if one of the points of intersection is chosen as center of inversion and any radius of inversion is used?

6. Prove that if two intersecting circles are orthogonal to a third circle, their points of intersection are inverse points with regard to the third circle.

7. If a circle with center C does not pass through the center of inversion O, and if OC cuts the circle in points P and Q, prove that $\frac{C'Q'}{C'P'} = \frac{OQ'}{OP'}$. In other words, C' is the harmonic conjugate of O with regard to P' and Q'.

8. If three circles intersect in a point, prove that four circles can be constructed tangent to all of them by inverting the figure, using the point of intersection as center of inversion.

9. By inverting a triangle with regard to any point as center of inversion, prove that the sum of the angles of a triangle is equal to two right angles.

10. Two circles intersect at O and P and their tangents at O meet the circles again at A and B. Show that the circle circumscribing the triangle OAB cuts OP produced at a point Q such that $OQ = 2 \cdot OP$. Prove by inverting the figure with regard to O.

11. If a transversal cuts two parallel lines it makes with them pairs of equal corresponding angles. Invert the figure composed of two parallels and a transversal, using any point as center of inversion, and obtain the *inverse* theorem, i.e., the corresponding theorem for the inverse figure.

12. Prove by inversion that there is in general one and only one circle through two given points orthogonal to a given circle. Invert with regard to one of the given points.

13. If points O, A, B, C lie on a circle, angles AOC and ABC are equal or supplementary. Invert with regard to O and derive the inverse theorem.

14. Prove that all circles which are tangent to one fixed circle and orthogonal to a second, will be tangent to a third.

15. An angle of fixed size rotates about a fixed point P, its sides cutting a fixed straight line in points Q and R. Prove that the circles circumscribed about the triangles PQR are tangent to a fixed circle.

16. Four bars are hinged together to form a linkwork as shown in Figure 128. Bars AB and DC are of equal length and so are AD and CB. The joints are at A, B, C

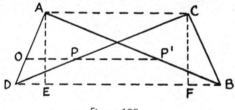

Figure 128

and D; AB and CD are crossed, but not fastened directly to one another. Points O, P and P' are located on DA, DC and AB, respectively, on any straight line parallel to AC and DB. Prove that $OP \cdot OP'$ is constant. Show how, by the addition of a radial bar, this four-bar inversor may be used to convert circular into rectilinear motion.

INDEX

(The numbers refer to pages)

245

A CATALOG OF SELECTED
DOVER BOOKS
IN SCIENCE AND MATHEMATICS

Astronomy

CHARIOTS FOR APOLLO: The NASA History of Manned Lunar Spacecraft to 1969, Courtney G. Brooks, James M. Grimwood, and Loyd S. Swenson, Jr. This illustrated history by a trio of experts is the definitive reference on the Apollo spacecraft and lunar modules. It traces the vehicles' design, development, and operation in space. More than 100 photographs and illustrations. 576pp. 6 3/4 x 9 1/4. 0-486-46756-2

EXPLORING THE MOON THROUGH BINOCULARS AND SMALL TELESCOPES, Ernest H. Cherrington, Jr. Informative, profusely illustrated guide to locating and identifying craters, rills, seas, mountains, other lunar features. Newly revised and updated with special section of new photos. Over 100 photos and diagrams. 240pp. 8 1/4 x 11. 0-486-24491-1

WHERE NO MAN HAS GONE BEFORE: A History of NASA's Apollo Lunar Expeditions, William David Compton. Introduction by Paul Dickson. This official NASA history traces behind-the-scenes conflicts and cooperation between scientists and engineers. The first half concerns preparations for the Moon landings, and the second half documents the flights that followed Apollo 11. 1989 edition. 432pp. 7 x 10. 0-486-47888-2

APOLLO EXPEDITIONS TO THE MOON: The NASA History, Edited by Edgar M. Cortright. Official NASA publication marks the 40th anniversary of the first lunar landing and features essays by project participants recalling engineering and administrative challenges. Accessible, jargon-free accounts, highlighted by numerous illustrations. 336pp. 8 3/8 x 10 7/8. 0-486-47175-6

ON MARS: Exploration of the Red Planet, 1958-1978--The NASA History, Edward Clinton Ezell and Linda Neuman Ezell. NASA's official history chronicles the start of our explorations of our planetary neighbor. It recounts cooperation among government, industry, and academia, and it features dozens of photos from Viking cameras. 560pp. 6 3/4 x 9 1/4. 0-486-46757-0

ARISTARCHUS OF SAMOS: The Ancient Copernicus, Sir Thomas Heath. Heath's history of astronomy ranges from Homer and Hesiod to Aristarchus and includes quotes from numerous thinkers, compilers, and scholasticists from Thales and Anaximander through Pythagoras, Plato, Aristotle, and Heraclides. 34 figures. 448pp. 5 3/8 x 8 1/2. 0-486-43886-4

AN INTRODUCTION TO CELESTIAL MECHANICS, Forest Ray Moulton. Classic text still unsurpassed in presentation of fundamental principles. Covers rectilinear motion, central forces, problems of two and three bodies, much more. Includes over 200 problems, some with answers. 437pp. 5 3/8 x 8 1/2. 0-486-64687-4

BEYOND THE ATMOSPHERE: Early Years of Space Science, Homer E. Newell. This exciting survey is the work of a top NASA administrator who chronicles technological advances, the relationship of space science to general science, and the space program's social, political, and economic contexts. 528pp. 6 3/4 x 9 1/4. 0-486-47464-X

STAR LORE: Myths, Legends, and Facts, William Tyler Olcott. Captivating retellings of the origins and histories of ancient star groups include Pegasus, Ursa Major, Pleiades, signs of the zodiac, and other constellations. "Classic." – Sky & Telescope. 58 illustrations. 544pp. 5 3/8 x 8 1/2. 0-486-43581-4

A COMPLETE MANUAL OF AMATEUR ASTRONOMY: Tools and Techniques for Astronomical Observations, P. Clay Sherrod with Thomas L. Koed. Concise, highly readable book discusses the selection, set-up, and maintenance of a telescope; amateur studies of the sun; lunar topography and occultations; and more. 124 figures. 26 halftones. 37 tables. 335pp. 6 1/2 x 9 1/4. 0-486-42820-6

Chemistry

MOLECULAR COLLISION THEORY, M. S. Child. This high-level monograph offers an analytical treatment of classical scattering by a central force, quantum scattering by a central force, elastic scattering phase shifts, and semi-classical elastic scattering. 1974 edition. 310pp. 5 3/8 x 8 1/2. 0-486-69437-2

HANDBOOK OF COMPUTATIONAL QUANTUM CHEMISTRY, David B. Cook. This comprehensive text provides upper-level undergraduates and graduate students with an accessible introduction to the implementation of quantum ideas in molecular modeling, exploring practical applications alongside theoretical explanations. 1998 edition. 832pp. 5 3/8 x 8 1/2. 0-486-44307-8

RADIOACTIVE SUBSTANCES, Marie Curie. The celebrated scientist's thesis, which directly preceded her 1903 Nobel Prize, discusses establishing atomic character of radioactivity; extraction from pitchblende of polonium and radium; isolation of pure radium chloride; more. 96pp. 5 3/8 x 8 1/2. 0-486-42550-9

CHEMICAL MAGIC, Leonard A. Ford. Classic guide provides intriguing entertainment while elucidating sound scientific principles, with more than 100 unusual stunts: cold fire, dust explosions, a nylon rope trick, a disappearing beaker, much more. 128pp. 5 3/8 x 8 1/2. 0-486-67628-5

ALCHEMY, E. J. Holmyard. Classic study by noted authority covers 2,000 years of alchemical history: religious, mystical overtones; apparatus; signs, symbols, and secret terms; advent of scientific method, much more. Illustrated. 320pp. 5 3/8 x 8 1/2.
0-486-26298-7

CHEMICAL KINETICS AND REACTION DYNAMICS, Paul L. Houston. This text teaches the principles underlying modern chemical kinetics in a clear, direct fashion, using several examples to enhance basic understanding. Solutions to selected problems. 2001 edition. 352pp. 8 3/8 x 11. 0-486-45334-0

PROBLEMS AND SOLUTIONS IN QUANTUM CHEMISTRY AND PHYSICS, Charles S. Johnson and Lee G. Pedersen. Unusually varied problems, with detailed solutions, cover of quantum mechanics, wave mechanics, angular momentum, molecular spectroscopy, scattering theory, more. 280 problems, plus 139 supplementary exercises. 430pp. 6 1/2 x 9 1/4. 0-486-65236-X

ELEMENTS OF CHEMISTRY, Antoine Lavoisier. Monumental classic by the founder of modern chemistry features first explicit statement of law of conservation of matter in chemical change, and more. Facsimile reprint of original (1790) Kerr translation. 539pp. 5 3/8 x 8 1/2. 0-486-64624-6

MAGNETISM AND TRANSITION METAL COMPLEXES, F. E. Mabbs and D. J. Machin. A detailed view of the calculation methods involved in the magnetic properties of transition metal complexes, this volume offers sufficient background for original work in the field. 1973 edition. 240pp. 5 3/8 x 8 1/2. 0-486-46284-6

GENERAL CHEMISTRY, Linus Pauling. Revised third edition of classic first-year text by Nobel laureate. Atomic and molecular structure, quantum mechanics, statistical mechanics, thermodynamics correlated with descriptive chemistry. Problems. 992pp. 5 3/8 x 8 1/2. 0-486-65622-5

ELECTROLYTE SOLUTIONS: Second Revised Edition, R. A. Robinson and R. H. Stokes. Classic text deals primarily with measurement, interpretation of conductance, chemical potential, and diffusion in electrolyte solutions. Detailed theoretical interpretations, plus extensive tables of thermodynamic and transport properties. 1970 edition. 590pp. 5 3/8 x 8 1/2. 0-486-42225-9

Engineering

FUNDAMENTALS OF ASTRODYNAMICS, Roger R. Bate, Donald D. Mueller, and Jerry E. White. Teaching text developed by U.S. Air Force Academy develops the basic two-body and n-body equations of motion; orbit determination; classical orbital elements, coordinate transformations; differential correction; more. 1971 edition. 455pp. 5 3/8 x 8 1/2. 0-486-60061-0

INTRODUCTION TO CONTINUUM MECHANICS FOR ENGINEERS: Revised Edition, Ray M. Bowen. This self-contained text introduces classical continuum models within a modern framework. Its numerous exercises illustrate the governing principles, linearizations, and other approximations that constitute classical continuum models. 2007 edition. 320pp. 6 1/8 x 9 1/4. 0-486-47460-7

ENGINEERING MECHANICS FOR STRUCTURES, Louis L. Bucciarelli. This text explores the mechanics of solids and statics as well as the strength of materials and elasticity theory. Its many design exercises encourage creative initiative and systems thinking. 2009 edition. 320pp. 6 1/8 x 9 1/4. 0-486-46855-0

FEEDBACK CONTROL THEORY, John C. Doyle, Bruce A. Francis and Allen R. Tannenbaum. This excellent introduction to feedback control system design offers a theoretical approach that captures the essential issues and can be applied to a wide range of practical problems. 1992 edition. 224pp. 6 1/2 x 9 1/4. 0-486-46933-6

THE FORCES OF MATTER, Michael Faraday. These lectures by a famous inventor offer an easy-to-understand introduction to the interactions of the universe's physical forces. Six essays explore gravitation, cohesion, chemical affinity, heat, magnetism, and electricity. 1993 edition. 96pp. 5 3/8 x 8 1/2. 0-486-47482-8

DYNAMICS, Lawrence E. Goodman and William H. Warner. Beginning engineering text introduces calculus of vectors, particle motion, dynamics of particle systems and plane rigid bodies, technical applications in plane motions, and more. Exercises and answers in every chapter. 619pp. 5 3/8 x 8 1/2. 0-486-42006-X

ADAPTIVE FILTERING PREDICTION AND CONTROL, Graham C. Goodwin and Kwai Sang Sin. This unified survey focuses on linear discrete-time systems and explores natural extensions to nonlinear systems. It emphasizes discrete-time systems, summarizing theoretical and practical aspects of a large class of adaptive algorithms. 1984 edition. 560pp. 6 1/2 x 9 1/4. 0-486-46932-8

INDUCTANCE CALCULATIONS, Frederick W. Grover. This authoritative reference enables the design of virtually every type of inductor. It features a single simple formula for each type of inductor, together with tables containing essential numerical factors. 1946 edition. 304pp. 5 3/8 x 8 1/2. 0-486-47440-2

THERMODYNAMICS: Foundations and Applications, Elias P. Gyftopoulos and Gian Paolo Beretta. Designed by two MIT professors, this authoritative text discusses basic concepts and applications in detail, emphasizing generality, definitions, and logical consistency. More than 300 solved problems cover realistic energy systems and processes. 800pp. 6 1/8 x 9 1/4. 0-486-43932-1

THE FINITE ELEMENT METHOD: Linear Static and Dynamic Finite Element Analysis, Thomas J. R. Hughes. Text for students without in-depth mathematical training, this text includes a comprehensive presentation and analysis of algorithms of time-dependent phenomena plus beam, plate, and shell theories. Solution guide available upon request. 672pp. 6 1/2 x 9 1/4. 0-486-41181-8

Browse over 9,000 books at www.doverpublications.com

HELICOPTER THEORY, Wayne Johnson. Monumental engineering text covers vertical flight, forward flight, performance, mathematics of rotating systems, rotary wing dynamics and aerodynamics, aeroelasticity, stability and control, stall, noise, and more. 189 illustrations. 1980 edition. 1089pp. 5 5/8 x 8 1/4. 0-486-68230-7

MATHEMATICAL HANDBOOK FOR SCIENTISTS AND ENGINEERS: Definitions, Theorems, and Formulas for Reference and Review, Granino A. Korn and Theresa M. Korn. Convenient access to information from every area of mathematics: Fourier transforms, Z transforms, linear and nonlinear programming, calculus of variations, random-process theory, special functions, combinatorial analysis, game theory, much more. 1152pp. 5 3/8 x 8 1/2. 0-486-41147-8

A HEAT TRANSFER TEXTBOOK: Fourth Edition, John H. Lienhard V and John H. Lienhard IV. This introduction to heat and mass transfer for engineering students features worked examples and end-of-chapter exercises. Worked examples and end-of-chapter exercises appear throughout the book, along with well-drawn, illuminating figures. 768pp. 7 x 9 1/4. 0-486-47931-5

BASIC ELECTRICITY, U.S. Bureau of Naval Personnel. Originally a training course; best nontechnical coverage. Topics include batteries, circuits, conductors, AC and DC, inductance and capacitance, generators, motors, transformers, amplifiers, etc. Many questions with answers. 349 illustrations. 1969 edition. 448pp. 6 1/2 x 9 1/4.
0-486-20973-3

BASIC ELECTRONICS, U.S. Bureau of Naval Personnel. Clear, well-illustrated introduction to electronic equipment covers numerous essential topics: electron tubes, semiconductors, electronic power supplies, tuned circuits, amplifiers, receivers, ranging and navigation systems, computers, antennas, more. 560 illustrations. 567pp. 6 1/2 x 9 1/4. 0-486-21076-6

BASIC WING AND AIRFOIL THEORY, Alan Pope. This self-contained treatment by a pioneer in the study of wind effects covers flow functions, airfoil construction and pressure distribution, finite and monoplane wings, and many other subjects. 1951 edition. 320pp. 5 3/8 x 8 1/2. 0-486-47188-8

SYNTHETIC FUELS, Ronald F. Probstein and R. Edwin Hicks. This unified presentation examines the methods and processes for converting coal, oil, shale, tar sands, and various forms of biomass into liquid, gaseous, and clean solid fuels. 1982 edition. 512pp. 6 1/8 x 9 1/4. 0-486-44977-7

THEORY OF ELASTIC STABILITY, Stephen P. Timoshenko and James M. Gere. Written by world-renowned authorities on mechanics, this classic ranges from theoretical explanations of 2- and 3-D stress and strain to practical applications such as torsion, bending, and thermal stress. 1961 edition. 560pp. 5 3/8 x 8 1/2. 0-486-47207-8

PRINCIPLES OF DIGITAL COMMUNICATION AND CODING, Andrew J. Viterbi and Jim K. Omura. This classic by two digital communications experts is geared toward students of communications theory and to designers of channels, links, terminals, modems, or networks used to transmit and receive digital messages. 1979 edition. 576pp. 6 1/8 x 9 1/4. 0-486-46901-8

LINEAR SYSTEM THEORY: The State Space Approach, Lotfi A. Zadeh and Charles A. Desoer. Written by two pioneers in the field, this exploration of the state space approach focuses on problems of stability and control, plus connections between this approach and classical techniques. 1963 edition. 656pp. 6 1/8 x 9 1/4.
0-486-46663-9

Browse over 9,000 books at www.doverpublications.com

Mathematics–Bestsellers

HANDBOOK OF MATHEMATICAL FUNCTIONS: with Formulas, Graphs, and Mathematical Tables, Edited by Milton Abramowitz and Irene A. Stegun. A classic resource for working with special functions, standard trig, and exponential logarithmic definitions and extensions, it features 29 sets of tables, some to as high as 20 places. 1046pp. 8 x 10 1/2. 0-486-61272-4

ABSTRACT AND CONCRETE CATEGORIES: The Joy of Cats, Jiri Adamek, Horst Herrlich, and George E. Strecker. This up-to-date introductory treatment employs category theory to explore the theory of structures. Its unique approach stresses concrete categories and presents a systematic view of factorization structures. Numerous examples. 1990 edition, updated 2004. 528pp. 6 1/8 x 9 1/4. 0-486-46934-4

MATHEMATICS: Its Content, Methods and Meaning, A. D. Aleksandrov, A. N. Kolmogorov, and M. A. Lavrent'ev. Major survey offers comprehensive, coherent discussions of analytic geometry, algebra, differential equations, calculus of variations, functions of a complex variable, prime numbers, linear and non-Euclidean geometry, topology, functional analysis, more. 1963 edition. 1120pp. 5 3/8 x 8 1/2. 0-486-40916-3

INTRODUCTION TO VECTORS AND TENSORS: Second Edition--Two Volumes Bound as One, Ray M. Bowen and C.-C. Wang. Convenient single-volume compilation of two texts offers both introduction and in-depth survey. Geared toward engineering and science students rather than mathematicians, it focuses on physics and engineering applications. 1976 edition. 560pp. 6 1/2 x 9 1/4. 0-486-46914-X

AN INTRODUCTION TO ORTHOGONAL POLYNOMIALS, Theodore S. Chihara. Concise introduction covers general elementary theory, including the representation theorem and distribution functions, continued fractions and chain sequences, the recurrence formula, special functions, and some specific systems. 1978 edition. 272pp. 5 3/8 x 8 1/2.
0-486-47929-3

ADVANCED MATHEMATICS FOR ENGINEERS AND SCIENTISTS, Paul DuChateau. This primary text and supplemental reference focuses on linear algebra, calculus, and ordinary differential equations. Additional topics include partial differential equations and approximation methods. Includes solved problems. 1992 edition. 400pp. 7 1/2 x 9 1/4. 0-486-47930-7

PARTIAL DIFFERENTIAL EQUATIONS FOR SCIENTISTS AND ENGINEERS, Stanley J. Farlow. Practical text shows how to formulate and solve partial differential equations. Coverage of diffusion-type problems, hyperbolic-type problems, elliptic-type problems, numerical and approximate methods. Solution guide available upon request. 1982 edition. 414pp. 6 1/8 x 9 1/4. 0-486-67620-X

VARIATIONAL PRINCIPLES AND FREE-BOUNDARY PROBLEMS, Avner Friedman. Advanced graduate-level text examines variational methods in partial differential equations and illustrates their applications to free-boundary problems. Features detailed statements of standard theory of elliptic and parabolic operators. 1982 edition. 720pp. 6 1/8 x 9 1/4. 0-486-47853-X

LINEAR ANALYSIS AND REPRESENTATION THEORY, Steven A. Gaal. Unified treatment covers topics from the theory of operators and operator algebras on Hilbert spaces; integration and representation theory for topological groups; and the theory of Lie algebras, Lie groups, and transform groups. 1973 edition. 704pp. 6 1/8 x 9 1/4.
0-486-47851-3

Browse over 9,000 books at www.doverpublications.com

A SURVEY OF INDUSTRIAL MATHEMATICS, Charles R. MacCluer. Students learn how to solve problems they'll encounter in their professional lives with this concise single-volume treatment. It employs MATLAB and other strategies to explore typical industrial problems. 2000 edition. 384pp. 5 3/8 x 8 1/2. 0-486-47702-9

NUMBER SYSTEMS AND THE FOUNDATIONS OF ANALYSIS, Elliott Mendelson. Geared toward undergraduate and beginning graduate students, this study explores natural numbers, integers, rational numbers, real numbers, and complex numbers. Numerous exercises and appendixes supplement the text. 1973 edition. 368pp. 5 3/8 x 8 1/2. 0-486-45792-3

A FIRST LOOK AT NUMERICAL FUNCTIONAL ANALYSIS, W. W. Sawyer. Text by renowned educator shows how problems in numerical analysis lead to concepts of functional analysis. Topics include Banach and Hilbert spaces, contraction mappings, convergence, differentiation and integration, and Euclidean space. 1978 edition. 208pp. 5 3/8 x 8 1/2. 0-486-47882-3

FRACTALS, CHAOS, POWER LAWS: Minutes from an Infinite Paradise, Manfred Schroeder. A fascinating exploration of the connections between chaos theory, physics, biology, and mathematics, this book abounds in award-winning computer graphics, optical illusions, and games that clarify memorable insights into self-similarity. 1992 edition. 448pp. 6 1/8 x 9 1/4. 0-486-47204-3

SET THEORY AND THE CONTINUUM PROBLEM, Raymond M. Smullyan and Melvin Fitting. A lucid, elegant, and complete survey of set theory, this three-part treatment explores axiomatic set theory, the consistency of the continuum hypothesis, and forcing and independence results. 1996 edition. 336pp. 6 x 9. 0-486-47484-4

DYNAMICAL SYSTEMS, Shlomo Sternberg. A pioneer in the field of dynamical systems discusses one-dimensional dynamics, differential equations, random walks, iterated function systems, symbolic dynamics, and Markov chains. Supplementary materials include PowerPoint slides and MATLAB exercises. 2010 edition. 272pp. 6 1/8 x 9 1/4. 0-486-47705-3

ORDINARY DIFFERENTIAL EQUATIONS, Morris Tenenbaum and Harry Pollard. Skillfully organized introductory text examines origin of differential equations, then defines basic terms and outlines general solution of a differential equation. Explores integrating factors; dilution and accretion problems; Laplace Transforms; Newton's Interpolation Formulas, more. 818pp. 5 3/8 x 8 1/2. 0-486-64940-7

MATROID THEORY, D. J. A. Welsh. Text by a noted expert describes standard examples and investigation results, using elementary proofs to develop basic matroid properties before advancing to a more sophisticated treatment. Includes numerous exercises. 1976 edition. 448pp. 5 3/8 x 8 1/2. 0-486-47439-9

THE CONCEPT OF A RIEMANN SURFACE, Hermann Weyl. This classic on the general history of functions combines function theory and geometry, forming the basis of the modern approach to analysis, geometry, and topology. 1955 edition. 208pp. 5 3/8 x 8 1/2. 0-486-47004-0

THE LAPLACE TRANSFORM, David Vernon Widder. This volume focuses on the Laplace and Stieltjes transforms, offering a highly theoretical treatment. Topics include fundamental formulas, the moment problem, monotonic functions, and Tauberian theorems. 1941 edition. 416pp. 5 3/8 x 8 1/2. 0-486-47755-X

Browse over 9,000 books at www.doverpublications.com

Mathematics–Logic and Problem Solving

PERPLEXING PUZZLES AND TANTALIZING TEASERS, Martin Gardner. Ninety-three riddles, mazes, illusions, tricky questions, word and picture puzzles, and other challenges offer hours of entertainment for youngsters. Filled with rib-tickling drawings. Solutions. 224pp. 5 3/8 x 8 1/2. 0-486-25637-5

MY BEST MATHEMATICAL AND LOGIC PUZZLES, Martin Gardner. The noted expert selects 70 of his favorite "short" puzzles. Includes The Returning Explorer, The Mutilated Chessboard, Scrambled Box Tops, and dozens more. Complete solutions included. 96pp. 5 3/8 x 8 1/2. 0-486-28152-3

THE LADY OR THE TIGER?: and Other Logic Puzzles, Raymond M. Smullyan. Created by a renowned puzzle master, these whimsically themed challenges involve paradoxes about probability, time, and change; metapuzzles; and self-referentiality. Nineteen chapters advance in difficulty from relatively simple to highly complex. 1982 edition. 240pp. 5 3/8 x 8 1/2. 0-486-47027-X

SATAN, CANTOR AND INFINITY: Mind-Boggling Puzzles, Raymond M. Smullyan. A renowned mathematician tells stories of knights and knaves in an entertaining look at the logical precepts behind infinity, probability, time, and change. Requires a strong background in mathematics. Complete solutions. 288pp. 5 3/8 x 8 1/2.
0-486-47036-9

THE RED BOOK OF MATHEMATICAL PROBLEMS, Kenneth S. Williams and Kenneth Hardy. Handy compilation of 100 practice problems, hints and solutions indispensable for students preparing for the William Lowell Putnam and other mathematical competitions. Preface to the First Edition. Sources. 1988 edition. 192pp. 5 3/8 x 8 1/2. 0-486-69415-1

KING ARTHUR IN SEARCH OF HIS DOG AND OTHER CURIOUS PUZZLES, Raymond M. Smullyan. This fanciful, original collection for readers of all ages features arithmetic puzzles, logic problems related to crime detection, and logic and arithmetic puzzles involving King Arthur and his Dogs of the Round Table. 160pp. 5 3/8 x 8 1/2.
0-486-47435-6

UNDECIDABLE THEORIES: Studies in Logic and the Foundation of Mathematics, Alfred Tarski in collaboration with Andrzej Mostowski and Raphael M. Robinson. This well-known book by the famed logician consists of three treatises: "A General Method in Proofs of Undecidability," "Undecidability and Essential Undecidability in Mathematics," and "Undecidability of the Elementary Theory of Groups." 1953 edition. 112pp. 5 3/8 x 8 1/2. 0-486-47703-7

LOGIC FOR MATHEMATICIANS, J. Barkley Rosser. Examination of essential topics and theorems assumes no background in logic. "Undoubtedly a major addition to the literature of mathematical logic." – *Bulletin of the American Mathematical Society.* 1978 edition. 592pp. 6 1/8 x 9 1/4. 0-486-46898-4

INTRODUCTION TO PROOF IN ABSTRACT MATHEMATICS, Andrew Wohlgemuth. This undergraduate text teaches students what constitutes an acceptable proof, and it develops their ability to do proofs of routine problems as well as those requiring creative insights. 1990 edition. 384pp. 6 1/2 x 9 1/4. 0-486-47854-8

FIRST COURSE IN MATHEMATICAL LOGIC, Patrick Suppes and Shirley Hill. Rigorous introduction is simple enough in presentation and context for wide range of students. Symbolizing sentences; logical inference; truth and validity; truth tables; terms, predicates, universal quantifiers; universal specification and laws of identity; more. 288pp. 5 3/8 x 8 1/2. 0-486-42259-3

Browse over 9,000 books at www.doverpublications.com

Mathematics–Algebra and Calculus

VECTOR CALCULUS, Peter Baxandall and Hans Liebeck. This introductory text offers a rigorous, comprehensive treatment. Classical theorems of vector calculus are amply illustrated with figures, worked examples, physical applications, and exercises with hints and answers. 1986 edition. 560pp. 5 3/8 x 8 1/2.   0-486-46620-5

ADVANCED CALCULUS: An Introduction to Classical Analysis, Louis Brand. A course in analysis that focuses on the functions of a real variable, this text introduces the basic concepts in their simplest setting and illustrates its teachings with numerous examples, theorems, and proofs. 1955 edition. 592pp. 5 3/8 x 8 1/2.   0-486-44548-8

ADVANCED CALCULUS, Avner Friedman. Intended for students who have already completed a one-year course in elementary calculus, this two-part treatment advances from functions of one variable to those of several variables. Solutions. 1971 edition. 432pp. 5 3/8 x 8 1/2.   0-486-45795-8

METHODS OF MATHEMATICS APPLIED TO CALCULUS, PROBABILITY, AND STATISTICS, Richard W. Hamming. This 4-part treatment begins with algebra and analytic geometry and proceeds to an exploration of the calculus of algebraic functions and transcendental functions and applications. 1985 edition. Includes 310 figures and 18 tables. 880pp. 6 1/2 x 9 1/4.   0-486-43945-3

BASIC ALGEBRA I: Second Edition, Nathan Jacobson. A classic text and standard reference for a generation, this volume covers all undergraduate algebra topics, including groups, rings, modules, Galois theory, polynomials, linear algebra, and associative algebra. 1985 edition. 528pp. 6 1/8 x 9 1/4.   0-486-47189-6

BASIC ALGEBRA II: Second Edition, Nathan Jacobson. This classic text and standard reference comprises all subjects of a first-year graduate-level course, including in-depth coverage of groups and polynomials and extensive use of categories and functors. 1989 edition. 704pp. 6 1/8 x 9 1/4.   0-486-47187-X

CALCULUS: An Intuitive and Physical Approach (Second Edition), Morris Kline. Application-oriented introduction relates the subject as closely as possible to science with explorations of the derivative; differentiation and integration of the powers of x; theorems on differentiation, antidifferentiation; the chain rule; trigonometric functions; more. Examples. 1967 edition. 960pp. 6 1/2 x 9 1/4.   0-486-40453-6

ABSTRACT ALGEBRA AND SOLUTION BY RADICALS, John E. Maxfield and Margaret W. Maxfield. Accessible advanced undergraduate-level text starts with groups, rings, fields, and polynomials and advances to Galois theory, radicals and roots of unity, and solution by radicals. Numerous examples, illustrations, exercises, appendixes. 1971 edition. 224pp. 6 1/8 x 9 1/4.   0-486-47723-1

AN INTRODUCTION TO THE THEORY OF LINEAR SPACES, Georgi E. Shilov. Translated by Richard A. Silverman. Introductory treatment offers a clear exposition of algebra, geometry, and analysis as parts of an integrated whole rather than separate subjects. Numerous examples illustrate many different fields, and problems include hints or answers. 1961 edition. 320pp. 5 3/8 x 8 1/2.   0-486-63070-6

LINEAR ALGEBRA, Georgi E. Shilov. Covers determinants, linear spaces, systems of linear equations, linear functions of a vector argument, coordinate transformations, the canonical form of the matrix of a linear operator, bilinear and quadratic forms, and more. 387pp. 5 3/8 x 8 1/2.   0-486-63518-X